Diffusion Processes, Jump Processes, and Stochastic Differential Equations

Diffusion Processes, Jump Processes, and Stochastic Differential Equations

Wojbor A. Woyczyński

Case Western Reserve University, USA

CRC Press
Taylor & Francis Group
Boca Raton London New York

CRC Press is an imprint of the
Taylor & Francis Group, an **informa** business

A CHAPMAN & HALL BOOK

First edition published 2022
by CRC Press
6000 Broken Sound Parkway NW, Suite 300, Boca Raton, FL 33487-2742

and by CRC Press
2 Park Square, Milton Park, Abingdon, Oxon, OX14 4RN

© 2022 Wojbor A. Woyczyński
CRC Press is an imprint of Taylor & Francis Group, an Informa business

No claim to original U.S. Government works

Library of Congress Cataloging-in-Publication Data
Names: Woyczyński, W. A. (Wojbor Andrzej), 1943- author.
Title: Diffusion processes, jump processes, and stochastic differential
equations / Wojbor A. Woyczyński, Case Western Reserve University, USA.
Description: First edition. | Boca Raton : Chapman & Hall/CRC Press, 2022.|
Includes bibliographical references and index.
Identifiers: LCCN 2021032255 (print) | LCCN 2021032256 (ebook) |
ISBN 9781032100678 (hardback) | ISBN 9781032107271 (paperback) |
ISBN 9781003216759 (ebook)
Subjects: LCSH: Diffusion processes. | Jump processes. |
Stochastic differential equations.
Classification: LCC QA274.75 .W69 2022 (print) | LCC QA274.75 (ebook) |
DDC 519.2/33—dc23
LC record available at https://lccn.loc.gov/2021032255
LC ebook record available at https://lccn.loc.gov/2021032256

ISBN-13: 978-1-03-210067-8 (hbk)
ISBN-13: 978-1-03-210727-1 (pbk)
ISBN-13: 978-1-00-321675-9 (ebk)

DOI: 10.1201/9781003216759

Typeset in Minion
by codeMantra

The children and I so want to dedicate the book to:

The Students of Prof. Wojbor A. Woyczyński

Prof. Paromita Banerjee, his last PhD graduate, and her husband, Prof. Anirban Mondal, his colleague at Case Western Reserve University, assisted with the posthumous preparations required for publication.

—Elizabeth H. Woyczyński

Contents

Author

Wojbor Andrzej Woyczyński was a mathematics and statistics professor, who was born and educated in Poland. He earned his M.Sc. in Electrical and Computer Engineering at Wroclaw University of Technology in 1966, and his Ph.D. in Mathematics, at the University of Wroclaw in 1968, when he was 23 years old. He spent most of his career teaching at Case Western Reserve University in Cleveland, Ohio, USA where he started working in 1982, when he was hired as chair of the Department of Mathematics and Statistics. He published over 160 papers and many books, including this, his 18th book, delving into a wide array of topics in mathematics. His research interests stretched from mainstream probability theory, to mathematical physics and turbulence theory, operations research and financial mathematics, to mathematical biology. In 1992 he published a monograph on "Random Series and Stochastic Integrals", co-written with Stanisław Kwapień. The paper "Lévy Flights in Evolutionary Ecology," co-written with two French mathematicians, Sylvie Méléard and Benjamin Jourdain, won the 2013 prize *La Recherche* for the best work in the field of mathematics. He published a number of works honoring the great mathematicians of preceding generations. Early in his career, in 1986, he was elected as a Fellow of the Institute of Mathematics. He served as an editorial board member of *Probability and Mathematical Statistics, Annals of Applied Probability*, and *Stochastic Processes and Their Applications*. This book was published posthumously, with the consent of his family.

CHAPTER 1

Random Variables, Vectors, Processes, and Fields

1.1 RANDOM VARIABLES, VECTORS, AND THEIR DISTRIBUTIONS—A GLOSSARY

1.1.1 Basic Concepts

We will denote by $(\Omega, \mathcal{F}, \mathbf{P})$ the *probability space (triple)* consisting of the *sample space* Ω, the σ-algebra $\mathcal{F}$ of subsets of Ω, and a probability measure $\mathbf{P}$ on $\mathcal{F}$. Elements $\omega \in \Omega$ will be called *sample points*, and elements of $\mathcal{F}$, *random events*. The probability measure $\mathbf{P}$ is assumed to be non-negative, countably additive, i.e., for a pairwise disjoint sequence $A_n, n - 1, 2, \ldots$, of random events,

$$\mathbf{P}\left(\bigcup_{n=1}^{\infty} A_n\right) = \sum_{n=1}^{\infty} \mathbf{P}(A_n),$$

and normalized, i.e., $\mathbf{P}(\Omega) = 1$.

A (real-valued) *random variable* is a mapping $X : \Omega \mapsto \mathbf{R}$ which is measurable, that is, a mapping such that, for any $x \in \mathbf{R}$, the set of sample points

$$\{\omega : X(\omega) \leq x\} = X^{-1}((-\infty, x])$$

is a random event, and its probability is well defined. In other words, for X to be a random variable, $\forall x \in \mathbf{R}$, the set $\{\omega : X(\omega) \leq x\}$ must be a member of the σ-field of random events $\mathcal{F}$. The set of all real-valued random variables will be denoted by $L_0(\Omega, \mathcal{F}, \mathbf{P}; \mathbf{R})$.

To each random variable X, we attach its *distribution*, μ_X, that is a measure on the Borel subsets $B \in \mathbf{R}$, defined by the equality

$$\mu_X(B) := \mathbf{P}(X^{-1}(B)) = \mathbf{P}(\omega : X(\omega) \in B) = \mathbf{P}(X \in B).$$

DOI: 10.1201/9781003216759-1

Recall that the Borel σ-field $\mathcal{B}$ of subsets of the real line is spanned by the half-lines $(-\infty, x], x \in \mathbf{R}$. Instead of using the distribution μ_X, it is often more convenient to operate with the *cumulative distribution function (CDF)* of X

$$F_X(x) := \mu_X((-\infty, x]) = \mathbf{P}(X \leq x).$$

Necessarily, F_X is non-decreasing, continuous on the right (because of countable additivity of μ), $F_X(-\infty) = 0$, and $F_X(+\infty) = 1$.

A non-negative function $f_X(x)$ on $\mathbf{R}$ such that

$$\mu_X(B) = \mathbf{P}(X \in B) = \int_B f_X(x)\, dx, \qquad \forall B \in \mathcal{B},$$

will be called the *probability density function (PDF)* of X. Then, necessarily, $\int_{\mathbf{R}} f_X(x)\, dx = 1$, and

$$F_X(x) = \int_{-\infty}^{x} f_X(z)\, dz, \qquad \text{and} \qquad F_X'(x) = f_X(x).$$

A mapping $\vec{X} : \Omega \mapsto \mathbf{R}^d$ is called a (*d*-dimensional) *random vector* if it is measurable, that is, for each Borel set $B \subset \mathbf{R}^d$, the set $\{\omega : \vec{X}(\omega) \in B\} \in \mathcal{F}$, and thus the probability measure $\mathbf{P}(\vec{X} \in B)$ is well defined. The set of all n-dimensional random vectors will be denoted by $L_0(\Omega, \mathcal{F}, \mathbf{P}; \mathbf{R}^d)$.

Equivalently,

$$\vec{X}(\omega) = (X_1(\omega), \ldots, X_d(\omega))$$

is a random vector if $\forall (x_1, \ldots, x_d) \in \mathbf{R}^d$, the set

$$\{\omega : X_1(\omega) \leq x_1, \ldots, X_d(\omega) \leq x_d\} \in \mathcal{F}.$$

In this case, the function of n variables,

$$F_{\vec{X}}(x_1, \ldots, x_d) = \mathbf{P}\big(\{\omega : X_1(\omega) \leq x_1, \ldots, X_d(\omega) \leq x_d\}\big)$$

is called the cumulative distribution function (CDF) of random vector $\vec{X}$.

A non-negative function $f_{\vec{X}}(\vec{x}) = f_{\vec{X}}(x_1, \ldots, x_d))$ on $\mathbf{R}^d$ such that

$$\mu_{\vec{X}}(B) := \mathbf{P}(\vec{X} \in B) = \int_B f_{\vec{X}}(x_1, \ldots, x_d)\, dx_1 \cdots dx_d, \qquad \forall B \in \mathcal{B}\backslash,$$

will be called the *probability density function (PDF)* of random vector $\vec{X}$. Then, necessarily, $\int_{\mathbf{R}^d} f_{\vec{X}}(\vec{x})\, dx_1 \cdots dx_d = 1$, and

$$F_{\vec{X}}(x_1, \ldots, x_d)) = \int_{-\infty}^{x_d} \cdots \int_{-\infty}^{x_1} f_{\vec{X}}(\vec{z})\, dz_1 \ldots dz_n,$$

and

$$f_{\vec{X}}(\vec{x}) = \frac{\partial^d}{\partial x_1 \ldots \partial x_d} F_{\vec{X}}(x_1, \ldots, x_d).$$

Components $X_1, \ldots, X_n$, of a random vector $\vec{X} = (X_1, \ldots, X_d)$ are said to be independent if, for every $\vec{x} \in \mathbf{R}^d$

$$F_{\vec{X}}(\vec{x}) = \mathbf{P}(X_1 \leq x_1, \ldots, X_d \leq x_d) = \prod_{k=1}^{d} \mathbf{P}(X_k \leq x_k) = F_{X_1}(x_1) \cdot \ldots \cdot F_{X_d}(x_d).$$

In the case of a random vector with a continuous distribution, independence of components means that the joint PDF is the product of marginal PDFs:

$$f_{\vec{X}}(\vec{x}) = f_{X_1}(x_1) \cdot \ldots \cdot f_{X_d}(x_d).$$

1.1.2 Absolutely Continuous, Discrete, Mixed, and Singular Probability Distributions

Probability distributions that possess densities are called *absolutely continuous*. On the opposite end of the spectrum are *discrete distributions* of the form

$$\mu(B) = \sum_{i=1}^{\infty} p_i \delta_{x_i}(B),$$

where $p_i \geq 0$, with $\mu(\mathbf{R}) = \sum_{i=1}^{\infty} p_i = 1$, which correspond to the random variables taking discrete values x_i with probability $p_i, i = 1, 2, \ldots$. Here,

$$\delta_x(B) = \left\{ \begin{array}{ll} 1, & \text{if } x \in B; \\ 0, & \text{if } x \notin B, \end{array} \right.$$

denotes the atomic *Dirac measure* concentrating mass 1 at point x.

Random variables such that

$$\mathbf{P}(X \in B) = \alpha \int_B f(x)\, dx + \beta \sum_{i=1}^{\infty} p_i \delta_{x_i}(B),$$

with $\alpha + \beta = 1, \alpha, \beta > 0$, and f and (p_i) normalized, are said to have a *mixed distribution*.

Some distributions are neither absolutely continuous, not discrete, nor their mixtures. An example here is the "devil's staircase" CDF which corresponds to the probability measure on the Cantor set, which has Lebesque measure 0, and the nontrivial fractional dimension. Such distributions are called *singular*. Of course, mixtures of all three, absolutely continuous, discrete, and singular distributions, can also be considered.

1.1.3 Characteristic Functions, Laplace Transforms, and Moment-Generating Functions

The probability distribution of a random variable X is also uniquely characterized by its (complex-valued) *characteristic function*

$$\varphi_X(u) = \mathbf{E}e^{iuX} = \int_{-\infty}^{\infty} e^{iux} \mu_X(dx) = \int_{-\infty}^{\infty} e^{iux} dF_X(x) = \int_{-\infty}^{\infty} e^{iux} f_X(x)\, dx,$$

the last part of the above formula is applicable in the case of absolutely continuous distributions. Given the characteristic function of X, its distribution can be recovered by the *inversion formula*. For, nonnegative-valued random variables, the characteristic function is often replaced by the real-valued *Laplace transform*,

$$\varphi_X(u) = \mathbf{E}e^{-uX} = \int_0^\infty e^{-ux}\mu_X(dx) = \int_0^\infty e^{-ux}dF_X(x) = \int_0^\infty e^{-ux}f_X(x)\,dx,$$

and for non-negative integer-valued X, by the *moment-generating function*

$$\psi_X(z) = \mathbf{E}z^X = \sum_{k=0}^\infty z^k\,\mathbf{P}(X=k).$$

In the latter case,
$$\psi_X(e^{iu}) = \varphi_X(u).$$

1.1.4 Examples

To establish the terminology and notation, a short list of standard examples of one- and multidimensional distributions is provided below.

Example 1.1 (Bernoulli distribution). *The random variable X takes two values, 1 and 0, with probabilities $p > 0$, and $1 - p$, respectively. The corresponding probability distribution is*

$$\mu_X = p\delta_1 + (1-p)\delta_0.$$

The characteristic and the moment-generating functions are, respectively,

$$\varphi_X(u) = pe^{iu} + (1-p), \qquad \psi_X(z) = pz + (1-p)$$

Example 1.2 (Binomial distribution). *In this case , given $1 > p > 0$,*

$$\mathbf{P}(X=k) = \binom{n}{k}p^k(1-p)^{n-k}, \qquad k = 0,1,2,\ldots,n,$$

with

$$\mu_X = \sum_{k=0}^n \binom{n}{k}p^k(1-p)^{n-k}\delta_k.$$

Example 1.3 (Poisson distribution). *For a $\lambda > 0$.*

$$\mathbf{P}(X=k) = e^{-\lambda}\frac{\lambda^k}{k!}, \qquad k = 0,1,2,\ldots,$$

with

$$\mu_X = \sum_{k=0}^\infty e^{-\lambda}\frac{\lambda^k}{k!}\delta_k.$$

Example 1.4 (Uniform distribution). *Random variable X uniformly distributed on a given interval $[a, b]$, has the PDF,*

$$f_X(x) = \begin{cases} (b-a)^{-1}, & \text{if } x \in [a, b]; \\ 0, & \text{if } x \notin [a, b]. \end{cases}$$

Example 1.5 (Exponential distribution). *Given the parameter $\lambda > 0$, the CDF in this case is described by the formula,*

$$F_X(x) = \begin{cases} 1 - e^{-\lambda x}, & \text{if } x \geq 0; \\ 0, & \text{if } x < 0. \end{cases}$$

The corresponding PDF is

$$f_X(x) = \begin{cases} \lambda e^{-\lambda x}, & \text{if } x \geq 0; \\ 0, & \text{if } x < 0. \end{cases}$$

Example 1.6 (Gaussian (normal) distribution). *There are two parameters, the mean $m \in \mathbf{R}$, and the variance σ^2. The distribution is usually referred to as the $N(m, \sigma^2)$-distribution, and its PDF is of the form*

$$\gamma_X(x) = \frac{1}{\sqrt{2\pi}\sigma} e^{-\frac{(x-m)^2}{2\sigma^2}}.$$

In this case, CDF cannot be evaluated explicitly in terms of elementary functions and, in the standard case, $m = 0, \gamma^2 = 1$, is usually denoted by $\Phi(x)$. You can think of it as a new (and important) special function (Figure 1.1).

Example 1.7 (Cauchy distribution). *The PDF here is of the form*

$$f_X(x) = \frac{1}{\pi(1 + x^2)}, \qquad -\infty < x < \infty.$$

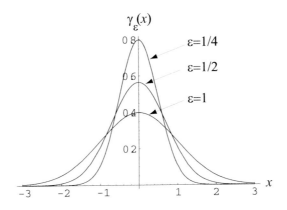

Figure 1.1   Gaussian PDFs with variances 1, 1/2, and 1/4.

An easy integration gives the CDF,

$$F_X(x) = \frac{1}{\pi}\left(\frac{\pi}{2} + \arctan x\right).$$

Example 1.8 (Multidimensional Gaussian distribution). *A Gaussian d-dimensional random vector $\vec{X} = (X_1, \ldots, X_d)^T$ (it will be more convenient to write this vector as a column) with independent components $X_i \sim N(m_i, \sigma_i^2)$, $i = 1, \ldots, d$, has the PDF*

$$f_{\vec{X}}(x_1, \ldots, x_d) = \frac{1}{(2\pi)^{d/2}\sigma_1\cdots\sigma_d}\exp\left(-\frac{(x_1 - m_1)^2}{2\sigma_1^2} - \cdots - \frac{(x_d - m_d)^2}{2\sigma_d^2}\right).$$

In the general case, when the components are not independent,

$$f_{\vec{X}}(x_1, \ldots, x_d) = \frac{1}{(2\pi)^{d/2}|\det\Sigma|^{1/2}}\exp\left(-\frac{1}{2}(\vec{x} - \vec{m})^T\,\Sigma^{-1}(\vec{x} - \vec{m})\right), \qquad (1.1)$$

where $\vec{m} = (m_1, \ldots, m_d) = (\mathbf{E}X_1, \ldots, \mathbf{E}X_d)$ is the mean vector, and $\Sigma = (\sigma_{ij}), i, j = 1, \ldots, d$, is the covariance matrix with $\sigma_{ij} = \mathbf{E}(X_i - m_i)(X_j - m_j)$, see Problem 1.3.2.

In the particular case of the planar random Gaussian vectors ($n = 2$) with mean zero, the above formula can be rewritten in the form

$$f_{\vec{X}}(x_1, x_2) = \frac{1}{2\pi\sigma_1\sigma_2\sqrt{1 - \rho^2}}\exp\left(-\frac{1}{2(1 - \rho^2)}\left(\frac{x_1^2}{\sigma_1^2} - 2\rho\frac{x_1 x_2}{\sigma_1\sigma_2} + \frac{x_2^2}{\sigma_2^2}\right)\right).$$

where $\rho = \sigma_{ij}/(\sigma_i\sigma_j)$ is the correlation coefficient between the first and second coordinates of the random vector $\vec{X} = (X_1, X_2)$.

An important observation here is that the probability distribution of a d-dimensional Gaussian random vector $\vec{X}$ is completely determined by the two-dimensional distributions of the collection of all pairs of components of $\vec{X}$. Indeed, all you need in formula (1.1) are the mean values of individual components, and the covariances of all the pairs of components.

Example 1.9 ("Devil's staircase" singular CDF). *The limit of the so-called "devil's staircase" CDFs shown in Figure 1.2 is an example of a CDF which, although continuous, and differentiable "almost everywhere", does not have a PDF.*

Observe that inside the interval [0,1] its derivative is 0 on the union of the infinite family of disjoint intervals whose lengths add up to 1. Indeed, as is clear from the construction displayed in Figure 1.2, this set has the linear measure

$$\lim_{n\to\infty}\left(\frac{1}{3} + 2\cdot\frac{1}{3^2} + + \cdots + 2^2\cdot\frac{1}{3^n}\right) = \frac{1}{3}\sum_{i=0}^{\infty}\left(\frac{2}{3}\right)^i = \frac{1}{3}\cdot\frac{1}{1 - 2/3} = 1,$$

in view of the formula for the sum of a geometric series. Thus, the integration of this derivative cannot possibly give a CDF that must grow from 0 to 1. Distributions

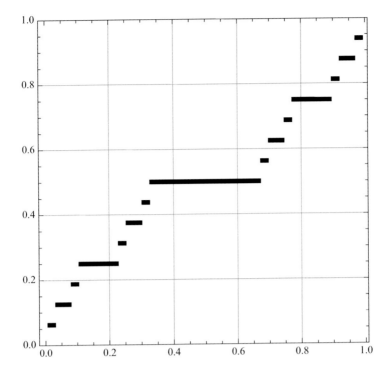

Figure 1.2 The construction of the singular "devil's staircase" c.d.f. $F_X(x)$. It continuously grows from 0, at $x = 0$, to 1, at $x = 1$, and yet it has no density; its derivative is equal to 0 on disjoint intervals whose lengths add up to 1.

of this type are called singular, and they arise in studies of fractal phenomena. One can prove that the set of points of increase of the limit "devil's staircase", i.e., the set of points on which the probability is concentrated, has a fractional dimension equal to $\ln 2 / \ln 3 = 0.6309\ldots$ [1]

1.2 LAW OF LARGE NUMBERS AND THE CENTRAL LIMIT THEOREM

In this section, we sketch two fundamental results of probability theory, the Law of large numbers and the Central Limit Theorem. Both deal with the analysis of the limit behavior of the averages

$$\bar{X}_n := \frac{X_1 + \cdots + X_n}{n}, \qquad n \to \infty,$$

of a sequence of independent, identically distributed random variables $X_1, X_2, \ldots,$ with common mean $\mu = \mathbf{E}X_i$, and common finite variance $\sigma^2 = \mathbf{E}(X_i - \mu)^2$.

---

[1] See, for example, M. Denker and W.A. Woyczynski, *Introductory Statistics and Random Phenomena: Uncertainty, Complexity and Chaotic Behavior in Engineering and Science*, Birkhäuser-Boston, 1998.

The (Weak) law of large numbers asserts that, for each $\epsilon > 0$,

$$\lim_{n \to \infty} \mathbf{P}(|\bar{X}_n - \mu| > \epsilon) = 0$$

or, in other words,

$$\lim_{n \to \infty} \bar{X}_n = \mu$$

in probability. The proof is immediate. First, observe that, in view of the independence and identical distribution of X_i's ,

$$\mathbf{E}(\bar{X}_n - \mu)^2 = \mathbf{E}\left(\frac{(X_1 - \mu) + \cdots + (X_n - \mu)}{n} \right)^2 = \frac{\sigma^2}{n}$$

so that, by Chebyshev's inequality

$$\mathbf{P}(|\bar{X}_n - \mu| > \epsilon) \leq \frac{\mathbf{E}(\bar{X}_n - \mu)^2}{\epsilon^2} = \frac{\sigma^2}{n\epsilon^2},$$

which yields the desired result.

The Central Limit Theorem gives us the limit behavior of the probability distribution of the rescaled fluctuations of the averages around the theoretical mean. More precisely, one starts with the observation that the random variables

$$\frac{\sqrt{n}}{\sigma}(\bar{X}_n - \mu), \qquad n = 1, 2, \ldots$$

all have mean zero and variance 1. Now it suffices to verify the shape of the limit of their characteristic functions. Taking into account the independence of X_i's we see that

$$\varphi_{\frac{\sqrt{n}}{\sigma}(\bar{X}_n - \mu)}(u) = \varphi_{X - \mu}^n\left(\frac{u}{n^{1/2}\sigma} \right).$$

However, by de l'Hospital rule (applied twice)

$$\lim_{n \to \infty} \log \varphi_{\frac{\sqrt{n}}{\sigma}(\bar{X}_n - \mu)}(u) = \lim_{n \to \infty} \frac{\log \varphi_{X - \mu}\left(\frac{u}{n^{1/2}\sigma} \right)}{n^{-1}}.$$

$$= \lim_{n \to \infty} \frac{\varphi_{X - \mu}^{-1}\left(\frac{u}{n^{1/2}\sigma} \right) \cdot \varphi_{X - \mu}'\left(\frac{u}{n^{1/2}\sigma} \right) \cdot \frac{-u}{2n^{3/2}\sigma}}{-n^{-2}}.$$

$$= \lim_{n \to \infty} \frac{\varphi_{X - \mu}''\left(\frac{u}{n^{1/2}\sigma} \right) \cdot \frac{-u^2}{2n^{3/2}\sigma^2}}{n^{-3/2}} = -\frac{u^2}{2},$$

so that the limit distribution of $\frac{\sqrt{n}}{\sigma}(\bar{X}_n - \mu)$, $n = 1, 2, \ldots$ is the standard Gaussian distribution with mean zero and variance one. Indeed, its characteristic function is

$$\int_{-\infty}^{\infty} e^{iux} \frac{e^{-x^2/2}}{\sqrt{2\pi}} dx = \int_{-\infty}^{\infty} \frac{e^{-\frac{1}{2}(x + iu)^2}}{\sqrt{2\pi}} dx \cdot e^{-u^2/2} = e^{-u^2/2}.$$

1.3 STOCHASTIC PROCESSES AND THEIR FINITE-DIMENSIONAL DISTRIBUTIONS

Let T be a subset of the real line (such as a finite interval or a half-line), and $(\Omega, \mathcal{F}, \mathbf{P})$ a probability space. A mapping

$$T \times \Omega \ni (t, \omega) \longmapsto X_t(\omega) \in \mathbf{R}, \qquad (1.2)$$

is called a stochastic process. Therefore, on the one hand, stochastic process is a function

$$T \ni t \longmapsto X_t \in L_0(\Omega, \mathcal{F}, \mathbf{P}; \mathbf{R}), \qquad (1.3)$$

with its values being random variables with the 1-D probability distributions

$$\mu_t(B) = \mathbf{P}(X_t \in B), \qquad B \in \mathcal{B}. \qquad (1.4)$$

On the other hand, the mapping

$$\Omega \ni \omega \longmapsto X.(\omega) \in \mathbf{R}^T \qquad (1.5)$$

has as its values real-valued functions on T which are called the *sample-paths (trajectories)* of process $(X_t, t \in T)$. This duality will be explored in some detail later on.

A more complete information about the process is given by its finite-dimensional distributions

$$\mu_{t_1, \dots, t_n}(B) = \mathbf{P}((X_{t_1}, \dots, X_{t_n}) \in B), \qquad B \in \mathcal{B}^n, \qquad (1.6)$$

$n = 1, 2, \dots; t_1, \dots, t_n \in T$, and the Kolmogorov's Consistency Theorem assures us that if we know all finite dimensional distribution, then we can determine (at least theoretically) the full infinite-dimensional probability distribution of the process on the space $\mathbf{R}^T$ of real functions on T (equipped with the natural sigma-field generated by the cylindrical sets).

Note that the finite-dimensional distributions of the process (X_t) satisfy the following *consistency condition*: for any $B \in \mathcal{B}^k$,

$$\mu_{t_1, \dots, t_k, t_{k+1}, \dots, t_n}(B \times \mathbf{R}^{n-k}) = \mu_{t_1, \dots, t_k}(B)$$

because

$$\mathbf{P}((X_{t_1}, \dots, X_{t_k}, X_{t_{k+1}}, \dots, X_{t_n}) \in B \times \mathbf{R}^{n-k}) = \mathbf{P}((X_{t_1}, \dots, X_{t_k}) \in B).$$

It is this fact which, through Komogorov's consistency theorem, guarantees the existence of a common extension of all finite-dimensional distributions to a probability measure, $\mu_{(X_t, t \in T)}$, on the space $\mathbf{R}^T$ of real functions f on T equipped with the σ-field generated by cylindrical sets with finite-dimensional bases. This means that if $B \in \mathcal{B}^k$, then

$$\mu_{(X_t, t \in T)}\left(\{f \in \mathbf{R}^T : (f(t_1), \dots, f(t_k)) \in B\}\right) = \mu_{t_1, \dots, t_k}(B).$$

Remark 1.1 (Almost sure properties of trajectories of a stochastic process). A study of a particular stochastic process often involves the question of almost sure (i.e., with probability 1) properties of its trajectories. Those questions are, essentially, questions about the measure $\mu_{(X_t, t \in T)}$ of the corresponding subset of the space $\mathbf{R}^T$ of all functions on T. Therefore, for example, if one can prove that $\mu^*_{(X_t, t \in T)}(C(T)) = 1$, then we can claim that, with probability 1, the trajectories of $X(t)$ are continuous. Notice, that we have cautiously added the asterisk to the measure μ. It denotes the outer measure of the set and may be needed here as, *a priori*, it is not obvious that $C(T)$ is a measurable subset of $\mathbf{R}^T$. We will see examples of this type of results in the following Chapters.

Remark 1.2 (Notation). In what follows we will write the time variable of the stochastic process either as a subscript, as in X_t, or as a variable, as in $X(t)$, the latter notation being useful in the case when an expression to be substituted for t is too bulky to fit in a subscript.

1.4 PROBLEMS AND EXERCISES

1.4.1. Prove right continuity of CDFs.

1.4.2. Show that the "devil's staircase" CDF generated the probability measure that is carried on the subset of $[0, 1]$ of fractional dimension $\log 2 / \log 3$. *Hint:* Use the definition of the fractional dimension in Section 2.5.

1.4.3. Verify that the linear (matrix) transformation $\vec{Y} = A\vec{X}$ of a standard Gaussian random vector has the PDF of the form (1.1) with the covariance matrix $\Sigma = AA^T$. *Hint:* Use the change of variables formula for integrals on $\mathbf{R}^d$, and find the Jacobian of the transformation.

1.4.4. The physical model that is usually mentioned to justify the introduction of (symmetric) Bernoulli distribution is "coin toss". Explain why it is relevant by considering the actual physical coin toss. Start with the coin of radius 1 and positioned heads-up at level 1. Then assume that it was flipped by imparting the initial vertical velocity v and angular velocity ω. Present the phase space (v, ω) picture of the outcomes, "heads-up" and "tails-up". Assuming a fixed error ϵ in controlling the initial values of v and ω, prove that as $v, \omega \to \infty$, the "probability" of either outcome converges to $1/2$.

CHAPTER 2

From Random Walk to Brownian Motion

2.1 SYMMETRIC RANDOM WALK; PARABOLIC RESCALING AND RELATED FOKKER-PLANCK EQUATIONS

2.1.1 Brownian Motion as Hydrodynamic Limit of Random Walks

Consider first a symmetric *random walk*,

$$X_n = \sum_{k=1}^{n} \xi_k, \quad (\xi_k) i.i.d, \quad \mathbf{P}(\xi_k = \pm 1) = \frac{1}{2}, \quad k = 1, 2, \ldots, n; n = 0, 1, , 2, \ldots,$$

on the one-dimensional lattice. Starting from the origin, the particle moves one step to the right, or left, with an equal probability 1/2 (see Figure 2.1). The consecutive steps are assumed to be statistically independent.

Now, to rescale it, the consecutive steps are accelerated and taken at the times

$$t_i = i\Delta t, \qquad i = 0, 1, 2, \ldots,$$

with the step size (lattice distance) Δx so that the set of possible positions of the particle is

$$x_k = k\Delta x, \qquad k = \ldots, -2, -1, 0, 1, 2, \ldots.$$

If $X(t, x)$ indicates whether site x is occupied ($X = 1$), or unoccupied ($X = 0$) at time t, and we denote by

$$f(t_i, x_k) = \boldsymbol{P}\{X(t_i, x_k) = 1\} \tag{2.1}$$

the probability that at time t_i the particle is located at site x_k then, clearly,

$$f(t_{i+1}, x_k) = \frac{1}{2} f(t_i, x_{k-1}) + \frac{1}{2} f(t_i, x_{k+1}). \tag{2.2}$$

DOI: 10.1201/9781003216759-2

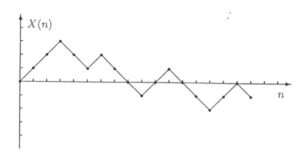

Figure 2.1 A sample path (trajectory) of the random walk X_n.

This equation can be rewritten as a system of difference equations for function f dependent on discrete time and space parameters,

$$f(t_{i+1}, x_k) - f(t_i, x_k) = \frac{1}{2}\left(f(t_i, x_{k-1}) + f(t_i, x_{x+1}) - 2f(t_i, x_k)\right), \qquad (2.3)$$

$i = 0, 1, 2, \ldots, k = \ldots, -2, -1, 0, 1, 2, \ldots$, with the initial condition $f(0, x) = \delta(x)$. Instead of directly solving the system (2.3), it is easier to notice that with *parabolic scaling*

$$\Delta t = (\Delta x)^2$$

we can rewrite (2.3) in the form

$$\frac{f(t_{i+1}, x_k) - f(t_i, x_k)}{\Delta t} = \frac{1}{2}\frac{f(t_i, x_{k-1}) + f(t_i, x_{k+1}) - 2f(t_i, x_k)}{(\Delta x)^2}, \qquad (2.4)$$

which, in the *hydrodynamic limit with parabolic scaling* $\Delta t = (\Delta x)^2 \to 0$, becomes the standard *linear diffusion (heat) equation*

$$\frac{\partial f}{\partial t} = \frac{1}{2}\frac{\partial^2 f}{\partial x^2}, \qquad (2.5)$$

for the time-dependent probability density function $f(t, x)$, depending on continuous time and space parameters, with the initial condition $f(0, x) = \delta(x)$.

The diffusion equation can be solved through the Fourier transform in the spatial variable. Indeed, denoting

$$\phi(t, \lambda) = (\mathcal{F}_x f)(t, \lambda) = \int_{-\infty}^{\infty} f(t, x)e^{-ix\lambda}\,dx$$

the diffusion equation is transformed into an ordinary linear differential equation (in the time variable t)

$$\frac{d\phi(t, \lambda)}{dt} = -\frac{1}{2}\lambda^2\phi(t, \lambda),$$

with an obvious exponential solution

$$\phi(t, \lambda) = \text{const} \cdot e^{-\lambda^2 t/2}.$$

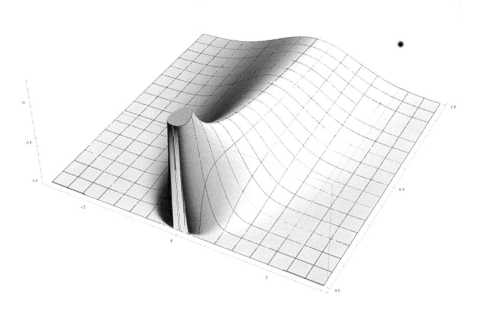

Figure 2.2 Evolution of the density $f(x,t)$ from the formula (2.1.1).

Taking the inverse Fourier transform,

$$(\mathcal{F}_\lambda^{-1}f)(t,x) = \frac{1}{2\pi}\int_{-\infty}^{\infty}\phi(t,\lambda)e^{ix\lambda}\,d\lambda,$$

and accounting for the initial condition $f(0,x) = \delta(x)$, gives the final solution of the diffusion equation

$$f(t,x) = \frac{1}{\sqrt{2\pi t}}e^{-x^2/(2t)}, \tag{2.6}$$

a Gaussian PDF with mean 0, and variance t. It is pictured in Figure 2.2 as a function of two variables, t and x.

Therefore, if we think about the Brownian motion $B(t), t > 0$, as the above parabolically rescaled limit of random walks, it is a Gaussian process with independent and stationary increments (because the independence and stationarity of the increments of random walk are preserved in the limit), with mean $\mathbf{E}B(t) = 0$, and variance $\mathbf{E}B^2(t) = t$. In addition, its covariance structure is clear: in view of the above properties, for $t_1 < t_2$,

$$\mathbf{E}B_{t_2}B_{t_1} = \mathbf{E}(B_{t_2} - B_{t_1})(B_{t_1} - B_0) + \mathbf{E}B_{t_1}^2 = t_1,$$

and for arbitrary $t, s > 0$,

$$\mathrm{Cov}\,(B_t B_s) = \mathbf{E}B_t B_s = t \wedge s \equiv \min(t, s). \tag{2.7}$$

Remark 2.1 (Asymmetric random walk and the diffusion with drift). If one permits asymmetric transition probabilities in the random walk, say, $p > 1/2$ for the particle moving to the right, and $q < 1/2$ for the particle moving to the left, then with the appropriate scaling in the hydrodynamic limit $\Delta t, \Delta x \to 0$, one obtains the general diffusion equation,

$$\frac{\partial f}{\partial t} + \delta \frac{\partial f}{\partial x} = \nu \frac{\partial^2 f}{\partial x^2} \tag{2.8}$$

(see, Problem 1). Physically, δ represents the *drift coefficient* and ν – the *viscosity (diffusion) coefficient*. The (3D versions of) the above arguments go back to the celebrated 1905–1906 papers of Albert Einstein and Marian Smoluchowski.

2.1.2 Brownian Motion via the Central Limit Theorem and the Invariance Principle

The above-mentioned approach to the derivation of Brownian motion through the densities and partial differential equations has its analogue in the context of the Central Limit Theorem.

Indeed, consider a sequence $\xi_n, n = 1, 2, \ldots$, of independent, identically distributed Bernoulli random variables with $\mathbf{P}(\xi_n = \pm 1) = 1/2$, and define, for a fixed $t = k/n \in [0, 1], k = 1, 2, \ldots, n$,

$$B_n(t) = n^{-1/2}(\xi_1 + \cdots + \xi_k).$$

The characteristic function

$$\phi_{B_n(t)}(\lambda) = \mathbf{E} e^{i\lambda B_n(t)} = \left(\mathbf{E} \exp\left[i\lambda \xi_1/\sqrt{n}\right]\right)^k = \cos^k(u/\sqrt{n}).$$

As $n, k \to \infty$, applying the l'Hospital rule to the variable n twice, we get

$$\lim_{n,k\to\infty} \log \phi_{B_n(t)}(\lambda) = \lim_{n,k\to\infty} k \log \cos \frac{\lambda}{\sqrt{n}} = \lim_{n,k\to\infty} \frac{k}{n} \frac{\log \cos(\lambda/\sqrt{n})}{1/n}$$

$$= t \lim_{n\to\infty} \frac{-\sin(\lambda/\sqrt{n})(-\lambda/2n^{3/2})}{-(1/n^2)\cos(\lambda/\sqrt{n})}$$

$$= t \lim_{n\to\infty} \frac{-\cos(\lambda/\sqrt{n})(\lambda^2/2n^{3/2})}{1/n^{3/2}} = -\frac{\lambda^2}{2}t,$$

which means that $\lim_{n\to\infty} \phi_{B_n(t)}(\lambda) = \exp[-t\lambda^2/2]$, so that, in law, the random variables $B_n(t)$ converge to a Gaussian random variable with mean zero and variance t, say $B(t)$, i.e., for all $x \in \mathbf{R}$,

$$\mathbf{P}\{B(t) \le x\} = \int_{-\infty}^{x} \frac{\exp(-y^2/(2t))}{\sqrt{2\pi t}}\, dy. \tag{2.9}$$

Interpolating the discrete-time process $B_n(t)$ introduced above to nonrational $t \in (0, 1)$ through the formula

$$B_n(t) = B_n(\lfloor nt \rfloor / n), \tag{2.10}$$

where $\lfloor x \rfloor$ is the greatest integer $\leq x$, defines, in the limit $n \to \infty$, the Brownian motion as a continuous time Gaussian stochastic process with prescribed mean and covariance structure . Indeed, $\mathbf{E}B_n(t) = 0$, and, for $0 \leq t < s \leq 1$,

$$\mathbf{E}\big(B_n(t)B_n(s)\big) = \mathbf{E}\left(\frac{\xi_1 + \cdots + \xi_{\lfloor nt \rfloor}}{\sqrt{n}} \cdot \frac{\xi_1 + \cdots + \xi_{\lfloor nt \rfloor} + \xi_{\lfloor nt \rfloor + 1} + \cdots + \xi_{\lfloor ns \rfloor}}{\sqrt{n}}\right)$$

$$= \frac{1}{n}\mathbf{E}(\xi_1^2 + \cdots + \xi_{\lfloor nt \rfloor}^2) = \frac{\lfloor nt \rfloor}{n} = \frac{nt + \epsilon_t}{nt} \cdot t \longrightarrow t, \qquad 0 < \epsilon_t \leq 1$$

as $n \to \infty$.

Since the finite-dimensional distributions of the random Gaussian vector are determined by their covariance structure, it can be shown that the finite-dimensional distributions of processes $B_n(t), t \in \mathbf{R}$, converge to the finite-dimensional distributions of the Brownian motion $B(t)$. This is a simple version of the celebrated *Invariance Principle* (see, e.g., Billingsley (1986), Theorem 37.8).

Thus, we arrive at the following economical, bare-bones, definition:

Definition 2.1. A process $B(t), t > 0$, with Gaussian finite-dimensional distributions, mean $\mathbf{E}B(t) = 0$, and covariance

$$\mathbf{E}\,B(t)B(s) = t \wedge s, \tag{2.11}$$

is called *Brownian motion*.

Figure 2.3 represents a few sample paths of a Brownian motion simulated from independent Gaussian increments. In the next section, we explicitly calculate

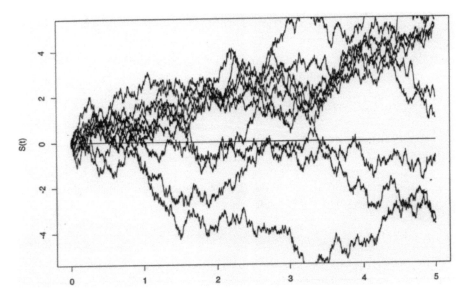

Figure 2.3   Sample paths of Brownian motion constructed from independent Gaussian increments.

those finite-dimensional distributions and establish some elementary properties of Brownian motion, which has been studied by the biologists[1], physicists, and mathematicians since the 1820s. However, only with the fundamental contribution of Norbert Wiener in 1926[2] has Brownian motion's mathematical studies become rigorous.

2.2 BASIC PROPERTIES OF BROWNIAN MOTION

Several basic properties of the Brownian motion follow directly from the definition. Indeed, we immediately get that,

2.2.1. Brownian motion starts at the origin: $B_0 = 0$,

2.2.2. Its variance grows linearly, $\mathbf{E}B_t^2 = t$, and, consequently,

$$\mathbf{P}(B_t \leq z) = \frac{1}{\sqrt{2\pi t}} \int_{-\infty}^{z} e^{-x^2/(2t)} \, dx. \tag{2.12}$$

2.2.3. The increments $B_{t_2} - B_{t_1}, 0 \leq t_1 < t_2$, of Brownian motion also have Gaussian distribution with mean zero and variance $t_2 - t_1$. To see why, observe that the increment

$$B_{t_2} - B_{t_1} = (-1, 1) \begin{pmatrix} B_{t_1} \\ B_{t_2} \end{pmatrix},$$

so it is a linear transformation of a Gaussian random vector and is thus Gaussian as well. Checking its variance using (2.11) gives

$$\mathbf{E}(B_{t_2} - B_{t_1})^2 = t_2 - 2(t_1 \wedge t_2) + t_1 = t_2 - t_1.$$

2.2.4. Brownian motion has stationary increments that is, for any $t_2 > t_1 \geq 0, \tau > 0$, the increments

$$B_{t_2} - B_{t_1}, \qquad \text{and} \qquad B_{t_2+\tau} - B_{t_1+\tau},$$

have identical Gaussian distributions with mean zero and variance $t_2 - t_1$. This property follows directly from Property 2.2.3.

2.2.5. B_t has orthogonal (uncorelated) increments over non-overlapping time intervals. Indeed, if $t_1 < t_2 < t_3$, then

$$\mathbf{E}(B_{t_2} - B_{t_1})(B_{t_3} - B_{t_2}) = t_2 \wedge t_3 - t_2 \wedge t_2 - t_1 \wedge t_3 + t_1 \wedge t_2$$
$$= t_2 - t_2 - t_1 + t_1 = 0.$$

---

[1]Robert Brown, a botanist, in 1827 observed under the microscope random movement of pollen particles on the surface of water, hence the name Brownian motion.

[2]Thus, in the mathematical literature, it is also common to call Brownian motion the Wiener process.

Since the finite-dimensional distributions of Brownian motion are assumed to be Gaussian, the orthogonality of the increments implies their statistical independence.

2.2.6. For any $t_0 > 0$, Brownian motion restarted at t_0 is also a Brownian motion, that is, the process $X_t = B_{t+t_0} - B_{t_0}, t > 0$, is also a Brownian motion. Indeed, it is Gaussian, the expectation is zero and

$$\mathbf{E}X_t X_s = \mathbf{E}(B_{t+t_0} - B_{t_0})(B_{s+t_0} - B_{t_0})$$
$$= (t + t_0) \wedge (s + t_0) - (t + t_0) \wedge t_0 - t_0 \wedge (s + t_0) + t_0 \wedge t_0$$
$$= t \wedge s.$$

2.2.7. Brownian motion is *self-similar* and has parabolic scaling, that is, for any $c > 0$, the process $X(t) = cB(t/c^2), t > 0$, is also a Brownian motion (see Figure 2.4). Here it also suffices to check the covariance structure:

$$\mathbf{E}X_s X_t = \mathbf{E}\big(cB(s/c^2) \cdot cB(t/c^2)\big) = c^2 \left(\frac{s}{c^2} \wedge \frac{t}{c^2}\right) = s \wedge t.$$

2.2.8. Inversion in time: The process $\{X(t) = tB(1/t), t > 0\}$, is also a Brownian motion. Thus, the behavior of $B(t)$ at infinity determines its behavior at

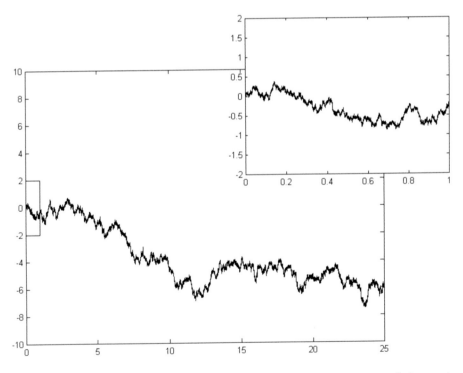

Figure 2.4 Self-similarity of the Brownian path: zooming in a piece of the trajectory produces, after rescaling, a trajectory with the same statistical properites.

zero, and vice versa. Again, checking the covariance structure, we get

$$\mathbf{E}X_s X_t = \mathbf{E}\big(sB(1/s) \cdot tB(1/t)\big) = st \left(\frac{1}{s} \wedge \frac{1}{t}\right) = t \wedge s.$$

2.2.9. Reflection in the time axis: The process $\{X(t) = -B(t), t > 0\}$, is also a Brownian motion. This is a direct consequence of the symmetry of the zero-mean Gaussian distribution, and we leave the verification to the reader.

2.2.10. The finite-dimensional distributions of the Brownian motion can be explicitly calculated:

$$\mathbf{P}(B_{t_1} \le z_1, \ldots, B_{t_n} \le z_n) = \int_{-\infty}^{z_1} \cdots \int_{-\infty}^{z_n} f_{t_1,\ldots,t_n}(x_1,\ldots.x_n)\, dx_1 \cdots dx_n, \tag{2.13}$$

with the PDFs, for $0 < t_1 < \cdots < t_n$,

$$f_{t_1,\ldots,t_n}(x_1,\ldots.x_n)$$
$$= \frac{e^{-x_1^2/(2t_1)}}{\sqrt{2\pi t_1}} \cdot \frac{e^{-(x_2-x_1)^2/(2(t_2-t_1))}}{\sqrt{2\pi(t_2 - t_1)}} \cdots \frac{e^{-(x_n-x_{n-1})^2/(2(t_n-t_{n-1}))}}{\sqrt{2\pi(t_n - t_{n-1})}}, \tag{2.14}$$

This result can be obtained from Properties 2.2.2, and 2.2.3, and the n-dimensional change-of-variables formula in the appropriate integral. See, Problem 2.5.2.

2.3   ALMOST SURE CONTINUITY OF SAMPLE PATHS

Brownian motion (has a version that) has continuous trajectories with probability 1. This result was proved by Norbert Wiener in 1926. In the 1960s, Zbigniew Ciesielski found an elegant proof of this fact using the orthonormal system of Haar wavelets[3] (see Figure 2.5) which are defined on the unit interval by the formula:

$$h_{k2^{-n}}(t) = \begin{cases} 2^{(n-1)/2}, & \text{for } (k-1)2^{-n} < t \le k2^{-n}; \\ -2^{(n-1)/2}, & \text{for } k2^{-n} < t \le (k+1)2^{-n}; \\ 0, & \text{elsewhere.} \end{cases} \tag{2.15}$$

for $n = 1, 2, \ldots$, and odd $k < 2^n$, with $h_0 \equiv 1$.

Theorem 2.1. *The process defined by the infinite random series*

$$X_t = \gamma_0 \int_0^t h_0 + \sum_{n \ge 1} \sum_{\text{odd } k < 2^n} \gamma_{k2^{-n}} \int_0^t h_{k2^{-n}}, \quad t > 0, \tag{2.16}$$

where $\gamma_{k2^{-n}}$ are independent, zero-mean Gaussian random variables with variance 1, is a Brownian motion. The series converges uniformly with probability 1, so that X_t is continuous with probability 1.

---

[3]Norbert Wiener used the trigonometric functions in his 1926 proof, which made the arguments more complicated.

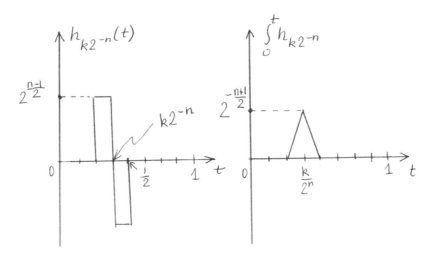

Figure 2.5 Example of a Haar wavelet $h_{k/2^n}(t)$ (a), and the corresponding Schauder tent function $\int_0^t h_{k/2^n}$ (b); here $2^n = 8$, and $k = 3$.

Before we present the formal proof it is worth considering the following (non-rigorous) intuition. Should, the derivative $\dot{B}_t$ of the Brownian motion exist (which it does not in the classical sense, but does in the distributional sense) then

$$\gamma_0 = \int_0^1 h_0 \dot{B}_t = B_1,$$

$$\gamma_{k2^{-n}} = \int_0^1 h_{k2^{-n}} \dot{B}_t = 2^{(n-1)/2} \left[B_{(k+1)/2^n} - 2B_{k/2^n} + B_{(k-1)/2^n} \right],$$

$$\mathbf{E}(\gamma_{k/2^n} \cdot \gamma_{j/2^n}) = \begin{cases} 0, & \text{if } k/2^n \neq j/2^n; \\ 1, & \text{if } k/2^n = j/2^n, \end{cases}$$

so that $\gamma_{k/2^n}, n = 1, 2, \ldots, k$ odd $< 2^n$, form a sequence of independent (orthogonal) zero mean, variance 1, Gaussian random variables, and one could expect an orthogonal expansion of the form

$$\dot{B}_t = \gamma_0 h_0 + \sum_{n \geq 1} \sum_{\text{odd } k < 2^n} \gamma_{k2^{-n}} h_{k2^{-n}}, \quad t > 0, \tag{2.17}$$

which justifies seeking the expansion of the Brownian motion itself given in the Theorem.

Proof. Each of the Schauder functions $\int_0^t f_{k2^{-n}}$ has a triangular shape (see, Figure 2.5). So the L_∞ norm of each term of the series (2.5)—a random number—

$$\varepsilon_n = \left\| \sum_{\text{odd } k < 2^n} \gamma_{k2^{-n}} \int_0^t h_{k2^{-n}} \right\|_\infty = 2^{-(n+1)/2} \max_{\text{odd } k < 2^n} |\gamma_{k2^{-n}}|. \tag{2.18}$$

Our job is now

(1) Prove that X_t has a.s. continuous sample path.

(2) Show that X_t is a Brownian motion,

To prove (1), we will have to demonstrate that the series (2.5) converges a.s. uniformly (Schauder functions are continuous). It will suffice to find a sequence of positive real numbers (a_n) such that, both

$$\sum_n a_n < \infty, \quad \text{and} \quad \sum_n \mathbf{P}(\varepsilon_n > a_n) < \infty, \tag{2.19}$$

seemingly two contradictory requirements. Should we accomplish this task, in view of the Borell-Cantelli Lemma, we would have

$$\mathbf{P}(\varepsilon_n \le a_n, \quad \text{for } n \text{ large enough}) = 1$$

or, equivalently,

$$\mathbf{P}(\varepsilon_n > a_n, \quad \text{infinitely often}) = 0,$$

and the proof of (1) would be complete.

Let's explore now the needed structure of the sequence (a_n).

$$\mathbf{P}(\varepsilon_n > a_n) = \mathbf{P}\left(2^{-(n+1)/2} \max_{\text{odd } k < 2^n} |\gamma_{k2^{-n}}| > a_n\right)$$

$$\le 2^{n-1} \cdot 2 \cdot \mathbf{P}(\gamma_0 > a_n 2^{(n+1)/2}) = 2^n \int_{a_n 2^{(n+1)/2}}^{\infty} \frac{e^{-x^2/2}}{\sqrt{2\pi}} dx$$

$$\le 2^n \frac{1}{a_n 2^{(n+1)/2}} e^{-a_n^2 2^{n+1}/2} \cdot \frac{1}{\sqrt{2\pi}} = \frac{1}{a_n} 2^{(n-1)/2} e^{-a_n^2 2^n} \cdot \frac{1}{\sqrt{2\pi}}$$

The structure of the right-hand side indicates that taking something like

$$a_n^2 = c2^{-n} \log 2^n$$

with a constant c would at least simplify the above expression. Indeed, we then get

$$\mathbf{P}(\varepsilon_n > a_n) \le \frac{1}{\sqrt{cn \log 2}} 2^{n-11/2} \cdot 2^{-cn} = \frac{1}{\sqrt{cn \log 2}} \frac{1}{2} \cdot 2^{n(1-c)}$$

Finally, taking $c > 1$, gives us a sequence (a_n), for which both series in (2.9) converge.

Now we can check that X_t is indeed a Brownian motion. Since the a.s. limit of Gaussian random variables is Gaussian the only thing remaining is to check the covariance structure of X_t. However, in view of Parseval's formula,

$$\mathbf{E}(X_t X_s) = \int_0^t h_0 \int_0^s h_0 + \sum_{n \ge 1} \sum_{\text{odd } k < 2^n} \int_0^t h_{k2^{-n}} \int_0^s h_{k2^{-n}} \tag{2.20}$$

$$= \int_0^1 I_{(0,t)} I_{(0,s)} = t \wedge s,$$

where I_A stands for the indicator function of A. $\qquad\square$

As the immediate corollary to the above theorem, we get the asymptotic behavior of the Brownian motion for large times.

Corollary 2.1. *(Law of large numbers)*

$$\mathbf{P}\left(\lim_{t\to\infty}\frac{B(t)}{t}=0\right)=1.$$

Indeed, in view of the a.s. continuity of the trajectories at zero, and the invariance property of the Brownian motion under the time inversion, we have

$$0=\lim_{t\to 0}B(t)=\lim_{t\to 0}tB(1/t).$$

2.4   NOWHERE DIFFERENTIABILITY OF BROWNIAN MOTION

The good news about the trajectories of the Brownian motion end at their continuity (a little more can be proved). The form of the covariance function immediately gives us a hint that $B(t)$ cannot be differentiable in the L_2 sense:

$$\mathbf{E}\left(\frac{B(t+h)-B(t)}{h}\right)^2=\frac{h}{h^2}\to\infty$$

as $0<h\to 0$. However, Wiener proved that a much stronger result holds true (the simple proof below is due to Dvoretzky, Erdös, and Kakutani, 1961).

Theorem 2.2. *Brownian motion is nowhere differentiable with probability 1. More precisely,*

$$\mathbf{P}(\dot{B}(t)\text{ exists for some }t)=0 \tag{2.21}$$

Proof. The differentiability of $B(t)$ at a point $s\in[0,1]$, implies the existence of an integer $N\geq 1$ such that, $|B(t)-B(s)|<N(t-s)$, as $t\searrow s$. More precisely, for sufficiently large n, and

$$j=\lfloor ns\rfloor+2, \lfloor ns\rfloor+3, \lfloor ns\rfloor+4,$$

we have

$$\left|B\left(\frac{j}{n}\right)-B\left(\frac{j-1}{n}\right)\right|<\frac{7N}{n},$$

because

$$\left|B\left(\frac{j}{n}\right)-B\left(\frac{j-1}{n}\right)\right|\leq\left|B\left(\frac{j}{n}\right)-B\left(s\right)\right|+\left|B\left(\frac{j-1}{n}\right)-B\left(s\right)\right|.$$

In other words,

$$\mathbf{P}(B'(t)\text{ exists for some }t\in(0,1))$$

$$\leq\mathbf{P}\left(\exists N\geq 1\,\exists s\,\exists m\geq 1\,\forall n\geq m\,\forall j=\lfloor ns\rfloor+2, \lfloor ns\rfloor+3, \lfloor ns\rfloor+4,\right.$$

$$\left| B\left(\frac{j}{n}\right) - B\left(\frac{j-1}{n}\right) \right| < \frac{7N}{n} \Big)$$

$$\leq \mathbf{P}\Big(\exists N \geq 1 \, \exists m \geq 1 \, \forall n \geq m \, \exists i = 2, \ldots, n+1 \, \forall j = i+1, i+2, i+3,$$

$$\left| B\left(\frac{j}{n}\right) - B\left(\frac{j-1}{n}\right) \right| < \frac{7N}{n} \Big)$$

$$\leq \mathbf{P}\Bigg(\bigcup_{N\geq 1} \bigcup_{m\geq 1} \bigcap_{n\geq m} \bigcup_{1<i\leq n+1} \bigcap_{i<j\leq i+3} \left(\left| B\left(\frac{j}{n}\right) - B\left(\frac{j-1}{n}\right) \right| < \frac{7N}{n} \right) \Bigg)$$

$$\leq \sum_{N=1}^{\infty} \mathbf{P}\Bigg(\liminf_{n\to\infty} \bigcup_{1<i\leq n+1} \bigcap_{i<j\leq i+3} \left(\left| B\left(\frac{j}{n}\right) - B\left(\frac{j-1}{n}\right) \right| < \frac{7N}{n} \right) \Bigg)$$

and, by Fatou Lemma,

$$\leq \sum_{N=1}^{\infty} \liminf_{n\to\infty} n \cdot \left(\mathbf{P}\left(\left| B\left(\frac{1}{n}\right) \right| < \frac{7N}{n} \right) \right)^3$$

$$= \sum_{N=1}^{\infty} \liminf_{n\to\infty} n \cdot \left(\mathbf{P}\left(|B(1)| < \frac{7N}{n} \cdot n^{1/2} \right) \right)^3 \leq \sum_{N=1}^{\infty} \liminf_{n\to\infty} n \cdot \left(\frac{7N}{\sqrt{2\pi}n^{1/2}} \right)^3 = 0$$
$\hfill \square$

Remark 2.2. Although $W(t) = \dot{B}(t)$, which is called the white noise, does not exist in the classical sense, it is an object that is commonly used in physics and engineering. It can be rigorously introduced in the framework of the theory of distributions (generalized functions), and we briefly describe below how this can be done. The *distributional derivative* of $B(t)$ exists, as for each ψ in the *Schwartz space* $\mathcal{D}(\mathbf{R})$, one can set the linear functional

$$\int \dot{B}(t)\,\psi(t)\,dt = \int \psi(t)\,dB(t),$$

where the *stochastic integral*[4] on the right is well defined; it is a Gaussian random variable with mean zero and variance $\int \psi^2(t)\,dt$. In other words, the mapping

$$L^2(\mathbf{R}) \ni \psi \mapsto \int \psi(t)\,dB(t) \in L^2_{Gauss}(\Omega)$$

is a linear isometry, and the basic rule of stochastic integration is $\mathbf{E}(dB(t))^2 = dt$!

The distributional-valued process $\dot{B}$, the *white noise*, is thus a delta-correlated generalized process since

$$\mathrm{Cov}\,(\dot{B}(t), \dot{B}(s)) = \delta(s - t).$$

Indeed, for any $\psi, \phi \in \mathcal{D}(\mathbf{R})$, the bilinear functional

$$\mathrm{Cov}\,(\dot{B}(t), \dot{B}(s))[\psi, \phi] = \mathbf{E}\left[\int \psi(t)\,dB(t) \int \phi(t)\,dB(t) \right]$$

$$= \int \psi(t)\phi(t)\,dt = \int\int \psi(t)\phi(s)\delta(s-t)\,dt\,ds.$$

---

[4]The detailed theory of stochastic integrals can be found in Chapter 6.

2.5 HITTING TIMES, AND OTHER SUBTLE PROPERTIES OF BROWNIAN MOTION

The time Brownian motion hits a certain level for the first time is called the *hitting time*, and it is the special case of a general theory of *stopping times* for stochastic processes. Formally, the hitting time T_b of level $b > 0$, a random quantity, is defined by the formula

$$T_b = \inf\{t : B(t) = b\}.$$

Note that the distribution of the hitting time is directly related to the distribution of the supremum of the Brownian motion. Indeed,

$$\mathbf{P}(T_b \geq t) = \mathbf{P}(\sup_{\tau \leq t} B(\tau) < b).$$

In view of the reflection principle

$$
\begin{aligned}
\mathbf{P}(T_b < t) &= \mathbf{P}(T_b < t, B(t) < b) + \mathbf{P}(T_b < t, B(t) > b) \\
&= 2\mathbf{P}(T_b < t, B(t) > b) = 2\mathbf{P}(B(t) > b)
\end{aligned}
$$

because of the strong Markov property (see 2) below),

$$= 2 \int_b^\infty \frac{e^{-x^2/2t}}{\sqrt{2\pi t}} = 2(1 - \Phi(b/\sqrt{t})),$$

where $\Phi(x)$ is the CDF of the standard $N(0, 1)$ Gaussian random variable. The the PDF of T_b is of the form

$$f_{T_b}(t) = \frac{b}{\sqrt{2\pi}t^{3/2}} e^{b^2/2t}, \tag{2.22}$$

for $b > 0$, and zero on the negative half-line.

Remark 2.3. The PDF (2.22) will appear later on in Chapter 5 as the special case of densities of stable Lévy jump processes; also, see Problem 2.6.3. It is often called the *inverse Gaussian PDF*.[5]

Here is a list of other important properties of Brownian motion (without proofs[6]):

(1) *Law of the iterated logarithm:* The law of large numbers for the Brownian motion discussed above demonstrated that $B(t)$ grows at the rate smaller than linear. The true rate of grows is precisely described by the following law of the iterated logarithm:

$$\mathbf{P}\left(\limsup_{t \searrow 0} \frac{|B(t)|}{\sqrt{2t \log \log t^{-1}}} = 1\right) = 1.$$

---

[5]For more information on the latter see ...

[6]For proofs see, e.g., ...

(2) *Strong Markov property:* Let $\mathcal{B}_t$ be the sub-σ-field of $\mathcal{F}$ spanned by $B_s, s \leq t$, and let $\tau : \Omega \to [0, \infty]$ be a Brownian stopping time, i.e., $\{\omega : \tau(\omega) < t\} \in \mathcal{B}_t \forall t$. Then, conditional on $\tau < \infty$,

$$Z(t) = B(t + \tau) - B(\tau), \quad t \geq 0,$$

is a Brownian motion independent of $B(t), t \leq \tau$.

(3) *Fractional dimension.* It can be proved that the fractional dimension of the Brownian sample path is

$$\dim\left[(t, B(t)) : t \in [0, 1]\right] = 3/2 \quad (= 2 - 1/2 \quad !!!)$$

Recall that the fractional dimension of a set $A \subset \mathbf{R}^d$ is defined by the formula

$$\dim[A] = \lim_{\epsilon \to 0} \frac{\log N_A(\epsilon)}{\log(1/\epsilon)}$$

where the coverage number $N_A(\epsilon)$ is the smallest number of d-dimensional balls or radius ϵ needed to cover A.

(4) *Vector Brownian motion.* A d-dimensional Brownian motion is defined as a random vector function

$$\vec{B}(t) = (B_1(t), \dots, B_d(t))$$

where $B_1(t), \dots, B_d(t)$ are independent one-dimensional Brownian motions. Note that the distributions of $\vec{B}(t)$ is invariant under d-dimensional rotations. Indeed, its joint PDF at time t is

$$f(\vec{x}) = f(x_1, \dots, x_d) = \frac{e^{-x_1^2/(2t)}}{\sqrt{2\pi t}} \cdot \dots \cdot \frac{e^{-x_d^2/(2t)}}{\sqrt{2\pi t}} = \frac{e^{-\|\vec{x}\|^2/(2t)}}{(2\pi t)^{d/2}},$$

where $\|\vec{x}\|$ stands for the Euclidean norm of the vector $\vec{x}$.

(5) *Recurrence and non-recurrence of Brownian motion in $\mathbf{R}^d$.* In dimensions 1 and 2, Brownian motion $\vec{B}(t)$ is recurrent, that is, for any open neighborhood of the origin, the probability of $B(t)$ returning to the neighborhood at some future time $t_0 > 0$ is equal to 1. That probability is equal to 0 in dimensions 3 and larger. The property follows from the analogous property for random walk in $\mathbf{R}^d$.

(6) Solution of the Dirichlet problem and Kakutani's theorem on hitting times.

2.6   PROBLEMS AND EXERCISES

2.6.1. Derive the general diffusion Equation (2.8) for the hydrodynamic limit of an asymmetric random walk. *Hint:* In addition to the parabolic scaling $\Delta x^2/\Delta t \to D$, you have to control the asymmetry asymptotically by assuming that $(q - p)/\Delta x \to A$, as $\Delta t \to 0$.

2.6.2. Derive the explicit formula (2.13–2.14) for the finite dimensional distributions of the Brownian motion. *Hint:* Write the random vector $\vec{X} = (B(t_1), \ldots, B(t_n))$ as a matrix A transformation of the random vector $\vec{Y} = (B(t_1) - B(0), B(t_2) - B(t_1), \ldots, B(t_n) - B(t_{n-1}))$ which has independent components with an easily expressed joint PDF. Then use the formula for changing variables in the multiple integral to calculate $\mathbf{P}(\vec{X} \in D) = \mathbf{P}(A\vec{Y} \in D)\mathbf{P}(\vec{Y} \in A^{-1}D)$, for an arbitrary domain $D \subset \mathbf{R}^n$.

2.6.3. Calculate the characteristic function of the PDF of the inverse Gaussian distribution (2.22) (the PDF of the hitting time of Brownian motion).

CHAPTER 3

Poisson Processes and Their Mixtures

3.1 WHY POISSON PROCESS?

Let us now move in the opposite direction from the continuous sample path Brownian motion model considered in the previous chapter, and consider a stochastic process $\{N_t, t \geq 0\}$, enjoying the following properties

(a) $N(t)$ takes values in the set of non-negative integers, and $N(0) = 0$.

(b) $N(t)$ has stationary increments, independent over non-overlapping intervals.

(c) $N(t)$ is nontrivial in the sense that, for each $t > 0$, we have $0 < \mathbf{P}(N(t) > 0) < 1$.

(d) $N(t)$ has only jumps of size 1.

The above set of properties seems like only a qualitative description of a process, but it turns out that they are restrictive enough to characterize the process completely in the quantitative sense.

Theorem 3.1. *If process $N(t), t > 0$, satisfies the above conditions (a)–(d), then it is a Poisson process, that is, there exists a constant $\mu > 0$, such that*

$$\mathbf{P}(N(t) = k) = e^{-\mu t}\frac{\mu^k t^k}{k!}, \qquad k = 0, 1, 2, \ldots. \tag{3.1}$$

Before we provide the proof of this result, some of the above conditions have to be restated to make them more precise.

(b')Stationarity of increments means that $\forall t > s, h > 0$

$$N(t) - N(s) \stackrel{d}{=} N(t + h) - N(s + h)$$

DOI: 10.1201/9781003216759-3

In particular, the increments are non-negative and the process trajectories are non-decreasing.

(d) Only jumps of size 1 are allowed means

$$\lim_{h \to 0} \frac{\mathbf{P}(N(h) - N(0) \geq 2)}{\mathbf{P}(N(h) - N(0) = 1)} = 0.$$

Proof. It suffices to prove that the moment generating function

$$\psi(z,t) := \sum_{n=0}^{\infty} z^n \mathbf{P}(N(t) = n) = e^{\mu t(z-1)}, \qquad |z| < 1. \tag{3.2}$$

Denote

$$\mathbf{P}(N(t) = n) = p_n(t).$$

Step 1. Since, in view of (a),(b), and (c),

$$p_0(t_1 + t_2) = \mathbf{P}(N(t_1 + t_2) = 0)$$
$$= \mathbf{P}(N(t_1) - N(0) = 0, N(t_1 + t_2) - N(t_1) = 0) = p_0(t_1) \cdot p_0(t_2),$$

there exists a $\mu > 0$, such that

$$p_0(t) = e^{-\mu t}. \tag{3.3}$$

Step 2. Since

$$\frac{\mathbf{P}(N(h) = 1)}{h} = \frac{1 - \mathbf{P}(N(h) = 0)}{h} - \frac{\mathbf{P}(N(h) \geq 2)}{h},$$

we have

$$\frac{\mathbf{P}(N(h) = 1)}{h} \left(1 + \frac{\mathbf{P}(N(h) \geq 2)}{\mathbf{P}(N(h) = 1)}\right) = \frac{1 - \mathbf{P}(N(h) = 0)}{h},$$

and for $h \to 0$, in view of (d), we have the second term in the parentheses converge to 0, and, in view of Step 1,

$$\lim_{h \to 0} \frac{p_1(h)}{h} = \lim_{h \to 0} \frac{\mathbf{P}(N(h) = 1)}{h} = \lim_{h \to 0} \frac{1 - e^{-\mu h}}{h} = \mu. \tag{3.4}$$

Step 3. Then, immediately from Step 2 and (d),

$$\lim_{h \to 0} \frac{\mathbf{P}(N(h) \geq 2)}{h} = \lim_{h \to 0} \frac{1 - \mathbf{P}(N(h) = 0)}{h} - \lim_{h \to 0} \frac{\mathbf{P}(N(h) = 1)}{h} = 0 \tag{3.5}$$

Step 4. In view of property (b), the moment generating function of $N(t)$ satisfies the following functional equation:

$$\psi(z, t + h) = \mathbf{E}(z^{N(t+h)}) = \mathbf{E}(z^{N(t+h) - N(h) + N(h) - N(0)}) = \psi(z, t) \cdot \psi(z, h)$$

so that

$$\frac{\psi(z, t+h) - \psi(z, t)}{h} = \psi(z, t) \cdot \frac{\psi(z, h) - 1}{h}$$

$$= \psi(z, t) \cdot \left(\frac{p_0(h) - 1}{h} + z \frac{p_1(h)}{h} + \sum_{n \geq 2} z^n \frac{p_n(h)}{h} \right)$$

As $h \to 0$, the left-hand side converges to $\partial \psi(z, t) / \partial t$, and, in view of Steps 1 and 2, the first two terms in the parenthesis on the right converge to $-\mu + z\mu$. For the third term, we have

$$\lim_{h \to 0} \sum_{n \geq 2} z^n \frac{p_n(h)}{h} \leq \lim_{h \to 0} \frac{\mathbf{P}(N(h) \geq 2)}{h} \sum_{n \geq 2} z^n = 0$$

in view of Step 3, and because the series converges for $|z| < 1$.

Therefore, the moment generating function must satisfy the ordinary differential equation

$$\frac{\partial \psi(z, t)}{\partial t} = \psi(z, t) \cdot (\mu(z - 1)) \tag{3.6}$$

with the initial condition $\psi(z, 0) = 1$. Its solution is the Poissonian moment generating function (3.2). $\qquad \square$

3.2 COVARIANCE STRUCTURE AND FINITE DIMENSIONAL DISTRIBUTIONS

Remembering that both the mean and variance of a Poisson distribution with parameter μ are equal to μ, we see that, in view condition (b), that the covariance, for $s < t$,

$$\mathrm{Cov}(N(t), N(s)) = \mathbf{E}(N(t) - \mu t)(N(s) - \mu s)$$
$$= \mathbf{E}[(N(t) - N(s)) + (N(s) - \mu s) + (\mu s - \mu t)] \cdot [(N(s) - \mu s)]$$
$$= \mu s.$$

Hence, for general t, s, we have

$$\mathrm{Cov}(N(t), N(s)) = \mu(t \wedge s), \tag{3.7}$$

so the covariance structure of the Poisson process is the same as that of the Brownian motion.

All the finite-dimensional distributions can be explicitly calculated. Similarly, various conditional probabilities can be explicitly evaluated. In particular, order statistics of a uniform distribution play here an important role. Let us proceed starting with the simplest two-dimensional cases.

Two-dimensional distributions. Consider $t_1 < t_2$. Then, because of (b), the process has nondecreasing trajectories, and

$$\mathbf{P}(N(t_1) = k_1, N(t_2) = k_2) = 0, \qquad \text{for} \qquad k_1 > k_2. \tag{3.8}$$

For $k_1 \leq k_2$,

$$
\begin{aligned}
\mathbf{P}(N(t_1) &= k_1, N(t_2) = k_2) \\
&= \mathbf{P}(N(t_2) - N(t_1) = k_2 - k_1) \cdot \mathbf{P}(N(t_1) = k_1) \\
&= \mathbf{P}(N(t_2 - t_1) = k_2 - k_1) \cdot \mathbf{P}(N(t_1) = k_1) \\
&= \frac{(\mu t_1)^{k_1} [\mu(t_2 - t_1)]^{k_2 - k_1} e^{\mu t_2}}{k_1!(k_2 - k_1)!}
\end{aligned}
\tag{3.9}
$$

Similarly, one can calculate the general n-dimensional distributions. It is clear that the distribution is supported by the simplex $\{(k_1, k_2, \ldots k_n) : k_1 \leq k_2 \leq \cdots \leq k_n\} \subset \mathbf{Z}_+^n$. However, see the chapter on general Lévy processes to see a simpler way to do it.

3.3   WAITING TIMES AND INTER-JUMP TIMES

The n-th waiting time is the random time when the process reaches the level n for the first time, i.e.,

$$
W_n = \min\{t : N(t) = n\}.
\tag{3.10}
$$

Theorem 3.2. *The n-th waiting time W_n of the Poisson process with parameter μ has the Gamma PDF*

$$
f_{W_n}(t) = e^{-\mu t} \frac{\mu^n t^{n-1}}{(n-1)!}, \qquad t \geq 0,
\tag{3.11}
$$

for $n = 1, 2, \ldots$.

Proof. Indeed, for $n = 1, 2, \ldots$, and $t \geq 0$, the cdf

$$
F_{W_n}(t) = \mathbf{P}(W_n \leq t) = \mathbf{P}(N(t) \geq n) = 1 - \sum_{k=0}^{n-1} e^{-\mu t} \frac{(\mu t)^k}{k!}
$$

Differentiating term by term, we obtain

$$
f_{W_n}(t) = F'_{W_n}(t) = -\sum_{k=1}^{n-1} e^{-\mu t} \frac{\mu k (\mu t)^{k-1}}{k!} + \sum_{k=0}^{n-1} e^{-\mu t} \frac{\mu(\mu t)^k}{k!} = e^{-\mu t} \frac{\mu^n t^{n-1}}{(n-1)!}
$$

$\square$

One easily checks that the characteristic function

$$
\varphi_{W_n}(u) = (1 - i\mu^{-1}u)^{-n}
\tag{3.12}
$$

with

$$
\mathbf{E}W_n = \mu^{-1}n, \qquad \text{and} \qquad \text{Var } W_n = \mu^{-2}n.
\tag{3.13}
$$

On the other hand, the random inter-jump times

$$
T_n = W_n - W_{n-1}, \qquad n = 1, 2, \ldots
\tag{3.14}
$$

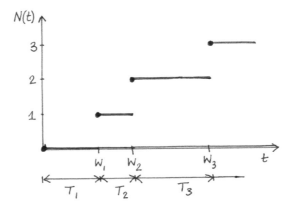

A sample path (trajectory) of a Poisson process; waiting times and inter-jump times are marked on the t-axis.

have a simpler structure as given in the next Theorem. Figure 3.1 represents one sample path (trajectory) of a Poisson process with the corresponding waiting times and inter-jump times.

Theorem 3.3. *Random variables $T_n, n = 1, 2, \ldots$, are independent, identically distributed with the exponential PDF with mean $\mathbf{E}T_n = \mu^{-1}$,*

$$f_{T_n}(t) = \mu e^{-\mu t}, \qquad t \geq 0. \tag{3.15}$$

Proof. (**sketch**). Observe, first that, for $n = 1$,

$$\mathbf{P}(T_1 > t) = \mathbf{P}(N(t) = 0) = e^{-\mu t}.$$

In general, we want to show that

$$\mathbf{P}(T_1 > t_1, T_2 > t_2, \ldots, T_n > t_n) = e^{-\mu t_1} \cdot e^{-\mu t_2} \cdots e^{-\mu t_n}.$$

But let's just check now the case $n = 2$:

$$\begin{aligned}
\mathbf{P}(T_1 > t_1, T_2 > t_2) &= \int_{t_1}^{\infty} \mathbf{P}(T_2 > t_2 | T_i = t) f_{T_1}(t) \, dt \\
&= \int_{t_1}^{\infty} \mathbf{P}\Big(N(t + t_2) - N(t) = 0\Big) f_{T_1}(t) \, dt \\
&= \int_{t_1}^{\infty} e^{-\mu t_2} \cdot \mu e^{-\mu t} \, dt = e^{-\mu t_2} \cdot e^{-\mu t_1}.
\end{aligned}$$

In general, one can proceed by induction, taking into account the fact that

$$\mathbf{P}(T_n > t | T_{n-1} = t_{n-1}, T_{n-2} = t_{n-2}, \ldots, T_1 = t_1)$$
$$\mathbf{P}\Big(N(t + t_1 + t_2 + \cdots + t_{n-1}) - N(t_1 + t_2 + \cdots + t_{n-1}) = 0\Big).$$

$\square$

Corollary 3.1. *Let (η_n) be a sequence of independent identically distributed random variables with the CDF $1 - e^{-\mu t}, t \geq 0$, and $\chi(t)$ be the Heaviside unit step function. Then*

$$N(t) = \sum_{n=1}^{\infty} \chi\left(t - \sum_{k=1}^{n} \eta_k\right), \qquad t \geq 0,$$

provides a representation of the Poisson process.

The distributions of jump waiting times conditioned on the present value of the Poisson process can also be explicitly determined. In particular, for $t_1 < t$

$$\mathbf{P}(W_1 = T_1 < t_1 | N(t) = 1) = \frac{\mathbf{P}(T_1 < t_1, N(t) = 1)}{\mathbf{P}(N(t) = 1)}$$

$$\frac{\mathbf{P}(N(t_1) - N(0) = 1, N(t) - N(t_1) = 0)}{\mathbf{P}(N(t) = 1)} = \frac{\mu t_1 e^{-\mu t_1} \cdot e^{-\mu(t - t_1)}}{\mu t e^{-\mu t}} = \frac{t_1}{t}.$$

So, the distribution is uniform on the interval $[0, t]$.

For the conditional joint distribution of the first two waiting time, we have

Theorem 3.4. *If $t_1 < t_2 < t$, then the conditional joint CDF*

$$F(t_1, t_2 | t) = \mathbf{P}(W_1 < t_1, W_2 < t_2 | N(t) = 2) == \left[t_1(t_2 - t_1) + \frac{t_1^2}{2} \right] \frac{2!}{t^2} \qquad (3.16)$$

and the conditional PDF

$$f(t_1, t_2 | t) = \frac{\partial^2}{\partial t_1 \partial t_2} F(t_1, t_2 | t) = \frac{2!}{t^2}. \qquad (3.17)$$

Proof. The random event considered in (3.16) can be decomposed into two disjoint events pictured in Figure 3.2

Thus (we check the formula for $\mu = 1$, the proof for general μ is identical)

$$\mathbf{P}(W_1 < t_1, W_2 < t_2 | N(t) = 2) = \frac{\mathbf{P}(W_1 < t_1, W_2 < t_2, N(t) = 2)}{\mathbf{P}(N(t) = 2)}$$

$$= \frac{\mathbf{P}(N(t_1) = 1, N(t_2) - N(t_1) = 1, N(t - t_2) = 0)}{\mathbf{P}(N(t) = 2)}$$

$$+ \frac{\mathbf{P}(N(t_1) = 2, N(t_2) - N(t_1) = 0, N(t - t_2) = 0)}{\mathbf{P}(N(t) = 2)}$$

$$= \left(e^{-t_1} \frac{t_1}{1!} \cdot e^{-(t_2 - t_1)} \frac{t_2 - t_1}{1!} \cdot e^{-(t - t_2)} \right) \Big/ \left(e^{-t} \frac{t^2}{2!} \right)$$

$$+ \left(e^{-t_1} \frac{t_1^2}{2!} \cdot e^{-(t_2 - t_1)} \cdot e^{-(t - t_2)} \right) \Big/ \left(e^{-t} \frac{t^2}{2!} \right)$$

$$= \left[t_1(t_2 - t_1) + \frac{t_1^2}{2} \right] \frac{2!}{t^2},$$

where both, the stationarity, and the independence of increments of the Poisson process have been utilized. □

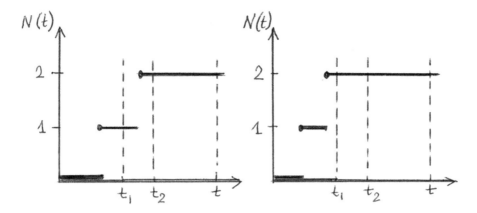

Figure 3.2 A decomposition of the event $\{W_1 < t_1, W_2 < t_2 | N(t) = 2\}$ into two disjoint events.

Remark 3.1 (Connection with order statistics). Similarly, one can establish that the joint probability

$$\mathbf{P}(W_1 < t_1, \ldots, W_n < t_n | N(t) = n)$$

has the uniform density $n!/t^n$, over the simplex $0 < t_1 < \cdots < t_n < t$ in $\mathbf{R}^n_+$ (see, Problem 3.4.2). Note that the distribution is exactly the ditribution of the n-th order statistics for the uniform distribution. In general, the PDF of n-th order statistics for i.i.d. random variables with an arbitrary density $f(x)$ is given by the formula

$$f(x_1, \ldots, x_n) = n! \prod_{i=1}^{n} f(x_i), \qquad \text{for} \quad x_1 < x_2 < \cdots < x_n.$$

In the same spirit, assuming that $N(t)$ is the Poisson process with rate $\mu > 0$, $0 < s < t$, and $0 \le k \le n$, we have

$$\mathbf{P}(N(s) = k | N(t) = n) = \frac{\left(e^{-\mu s}(\mu s)^k / k!\right) \cdot \left(e^{-\mu(t-s)}(\mu(t-s))^{n-k}/(n-k)!\right)}{\left(e^{-\mu t}(\mu t)^n / n!\right)}$$

$$= \binom{n}{k} \left(\frac{s}{t}\right)^k \left(1 - \frac{s}{t}\right)^{n-k}.$$

Remark 3.2 (Applications to reliability theory). Poisson processes are a standard tool in reliability theory (think about a light bulb that burns out and has to be replaced after a random exponentially distributed time). In this context, several random quantities are of interest, including current life δ_t, and excess life γ_t, which are informally defined in Figure 3.3.

The distributions of current life and excess life are easily determined: for a fixed $t > 0$,

$$1 - F_{\gamma_t}(s) = \mathbf{P}(\gamma_t > s) = \mathbf{P}(N(t + s) - N(t) = 0) = e^{-\mu s},$$

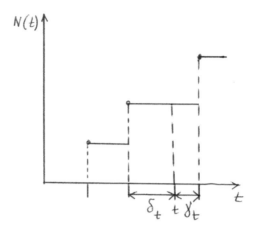

Figure 3.3 Current life, δ_t, and excess life ,γ_t, for a Poisson process.

and, for $s < t$,

$$F_{\delta_t}(s) = \mathbf{P}(\delta_t \leq s) = \mathbf{P}(N(t) - N(t-s) = 0) = 1 - e^{-\mu s}.$$

For $s \geq t$, obviously, $F_{\delta_t}(s) = 1$.

3.4   EXTENSIONS AND GENERALIZATIONS

(a) *A non-homogeneous process (non-stationary increments)*. The increments are independent but

$$\mathbf{P}(N(t) = n) = e^{-\mu(t)} \frac{\mu^n(t)}{n!},$$

where $\mu(t)$ is a non-negative and nondecreasing function.

(b) *Generalized Poisson process*. Integer valued with stationary and independent increments, but the jumps can be of any positive integer size. The characteristic function

$$\varphi_{N(t)}(u) = e^{\nu t[\phi(u)-1]},$$

where ν is a positive constant, and

$$\phi(u) = \sum_{k=1}^{\infty} p_k e^{iku}.$$

The non-negative constants p_k are "probabilities" of jumps of size k. More about this model in the next chapter.

(c) *Compound Poisson process*.

$$X(t) = \sum_{n=1}^{N(t)} Y_n,$$

where $N(t)$ is a Poisson process independent of the sequences (Y_n) of independent identically distributed random variables with characteristic function $\varphi_Y(u)$. Its characteristic function

$$\varphi_{X(t)}(u) = e^{\mu t[\varphi_Y(u)-1]}.$$

Indeed,

$$\mathbf{E}\exp\left(iu\sum_{n=1}^{N(t)}Y_n,\right) = \sum_{m=0}^{\infty}\mathbf{E}\left(\exp(iu\sum_{n=1}^{N(t)}Y_n)\;\Big|\;N(t)=m\right)\cdot\mathbf{P}(N(t)=m)$$

$$= \sum_{m=0}^{\infty}\varphi_Y^m(u)e^{-\mu t}\frac{(\mu t)^m}{m!} = e^{\mu t(\varphi_Y(u)-1)}.$$

3.5 FRACTIONAL POISSON PROCESSES (fPp)

3.5.1 FPp Interarrival Time

The fractional Poisson process $N_\nu(t)$, $0 < \nu \le 1$, $t > 0$, was defined in Repin and Saichev (2000) via the following formula for the Laplace transform of the p.d.f $\psi_\nu(t)$ of its i.i.d. interarrival times T_i, $i = 1, 2, \ldots$:

$$\{\mathsf{L}\psi_\nu(t)\}(\lambda) \equiv \widetilde{\psi}_\nu(\lambda) \equiv \int_0^{\infty} e^{-\lambda t}\psi_\nu(t)dt = \frac{\mu}{\mu + \lambda^\nu}, \tag{3.18}$$

where $\mu > 0$ is a parameter. For $\nu = 1$, the above transform coincides with the Laplace transform

$$\widetilde{\psi}_1(\lambda) = \frac{\mu}{\mu + \lambda}.$$

of the exponential interarrival time density of the ordinary Poisson process with parameter $\mu = \mathbf{E}N_1(1)$.

Using the inverse Laplace transform, the above-cited authors derived the *singular integral equation* for $\psi_\nu(t)$:

$$\psi_\nu(t) + \frac{\mu}{\Gamma(\nu)}\int_0^t \psi_\nu(\tau)\frac{d\tau}{[\mu(t-\tau)]^{1-\nu}} = \frac{\mu^\nu}{\Gamma(\nu)}t^{\nu-1},$$

which is equivalent to the *fractional differential equation*,

$$_0D_t^\nu\psi_\nu(t) + \mu\psi_\nu(t) = \delta(t),$$

where the Liouville derivative operator $_0D_t^\nu = d^\nu/dt^\nu$ (see, e.g., Kilbas, Srivastava and Trujillo, 2006) is defined via the formula

$$_0D_t^\nu\psi_\nu(\tau) = \frac{1}{\Gamma(1-\nu)}\frac{d}{dt}\int_0^t \psi_\nu(\tau)\frac{d\tau}{[\mu(t-\tau)]^{1-\nu}}.$$

These characterizations permitted them to obtain the following integral representation for the p.d.f. $\psi_\nu(t)$,

$$\psi_\nu(t) = \frac{1}{t} \int_0^\infty e^{-x} \phi_\nu(\mu t/x) dx, \tag{3.19}$$

where

$$\phi_\nu(\xi) = \frac{\sin(\nu\pi)}{\pi[\xi^\nu + \xi^{-\nu} + 2\cos(\nu\pi)]},$$

and demonstrate that the tail probability distribution of the waiting time T is of the form

$$\mathbf{P}(T > t) = \int_t^\infty \psi_\nu(\tau)\, d\tau = E_\nu(-\mu t^\nu), \tag{3.20}$$

where

$$E_\nu(z) = \sum_{n=0}^\infty \frac{z^n}{\Gamma(\nu n + 1)} \tag{3.21}$$

is the Mittag-Leffler function (see, e.g., Kilbas, Srivastava and Trujillo (2006)).

Remark 3.3. *Observe that the Mittag–Leffler function is a fractional generalization of the standard exponential function* $\exp(z)$*; indeed* $E_1(z) = \exp(z)$*. It has been widely used to describe probability distributions appearing in finance and economics, anomalous diffusion, transport of charge carriers in semiconductors, and light propagation through random media (Piryatinska, Saichev and Woyczynski, 2005; Uchaikin and Zolotarev, 1999).*

In view of (3.20–3.21), the interarrival time density for the fractional Poisson process can be easily shown to be

$$\psi_\nu(t) = \mu t^{\nu-1} E_{\nu,\nu}(-\mu t^\nu), \qquad t \geq 0, \tag{3.22}$$

where

$$E_{\alpha,\beta}(z) = \sum_{n=0}^\infty \frac{z^n}{\Gamma(\alpha n + \beta)}$$

is the generalized, two-parameter Mittag-Leffler function. Also, the above information automatically gives the p.d.f.

$$f_n^\nu(t) = \mu^n \nu \frac{t^{\nu n-1}}{(n-1)!} E_\nu^{(n)}\big(-\mu t^\nu\big), \tag{3.23}$$

of the n-the arrival time, $A_n = T - 1 + \cdots + T_n$, because, obviously, its Laplace transform,

$$\mathsf{L}\{f_n^\nu(t)\} = \frac{\mu^n}{(\mu + \lambda^\nu)^n}.$$

As $\nu \to 1$, the above distribution converges to the classical Erlang distribution.

Example 3.1. *For some values of ν, the p.d.f. of the interarrrival times can be calculated more explicitly. In particular, consider*

$$\psi_{1/2}(t) = \mu t^{1/2-1} E_{1/2,1/2}\left(-\mu t^{1/2}\right), \qquad t \geq 0,$$

where

$$E_{1/2,1/2}\left(-z\right) = \sum_{n=0}^{\infty} \frac{(-z)^n}{\Gamma\left(\frac{n}{2}+\frac{1}{2}\right)} = \frac{1}{\sqrt{\pi}} - z E_{1/2,1}(-z). \qquad (3.24)$$

Using the identity,

$$E_{1/2,1}(-z) = e^{z^2}\mathrm{Erfc}(z),$$

where

$$\mathrm{Erfc}(t) = \frac{2}{\sqrt{\pi}} \int_z^{\infty} e^{-u^2} du,$$

is the complementary error function, we obtain,

$$\psi_{1/2}(t) = \mu t^{-1/2}\left(\frac{1}{\sqrt{\pi}} - \mu t^{1/2} e^{\left(\mu t^{1/2}\right)^2}\mathrm{Erfc}(\mu\sqrt{t})\right)$$

$$= \frac{\mu}{\sqrt{\pi t}} - \mu^2 e^{\mu^2 t}\mathrm{Erfc}(\mu\sqrt{t}), \qquad t \geq 0. \qquad (3.25)$$

In another approach to the study of $N_\nu(t)$, Jumarie (2001), and Laskin (2003), used the fractional Kolmogorov–Feller-type differential equation system

$$_0D_t^\nu P_n^\nu(t) = \mu[P_{n-1}^\nu(t) - P_n^\nu(t)] + \delta_{n0}\frac{t^{-\nu}}{\Gamma(1-\nu)}, \qquad n = 1, 2, \ldots, \qquad (3.26)$$

to characterize the 1-D probability distributions $P_n^\nu(t) = \mathbf{P}\left(N_\nu(t) = n\right)$. The solutions of the above system of Equation (3.26) can be calculated to be

$$G_\nu(u,t) \equiv \mathbf{E}u^{N_\nu(t)} = E_\nu\left(\mu t^\nu(u-1)\right). \qquad (3.27)$$

Hence, expanding $G_\nu(u,t)$ over u, and rearranging (3.27)), we find

$$P_n^\nu(t) = \frac{(-z)^n}{n!}\frac{d^n}{dz^n}E_\nu(z)\bigg|_{z=-\mu t^\nu} = \frac{(\mu t^\nu)^n}{n!}\sum_{k=0}^{\infty}\frac{(k+n)!}{k!}\frac{(-\mu t^\nu)^k}{\Gamma(\nu(k+n)+1)}. \qquad (3.28)$$

Equivalently, one can show (see, Laskin, 2003) that the moment generating function (MGF) of the fractional Poisson process $N_\nu(t)$ is of the form

$$M_\nu(s,t) \equiv \mathbf{E}\,e^{-sN_\nu(t)} = \sum_{m=0}^{\infty}\frac{[\mu t^\nu\left(e^{-s}-1\right)]^m}{\Gamma(\nu m+1)}, \qquad (3.29)$$

which permits calculation (see, Table 3.1) of the fPp's moments via the usual formula,

$$\mathbf{E}\left[N_\nu(t)\right]^k = (-1)^k\frac{\partial^k}{\partial s^k}M_\nu(s,t)\bigg|_{s=0}.$$

TABLE 3.1 Properties of fPp compared with those of the Poisson process.

	Poisson Process ($\nu = 1$)	Fractional Poisson Process ($\nu < 1$)
$P_0(t)$	$e^{-\mu t}$	$E_\nu(-\mu t^\nu)$
$\psi(t)$	$\mu e^{-\mu t}$	$\mu t^{\nu-1} E_{\nu,\nu}(-\mu t^\nu)$
$P_n(t)$	$\frac{(\mu t)^n}{n!} e^{-\mu t}$	$\frac{(\mu t^\nu)^n}{n!} \sum_{k=0}^{\infty} \frac{(k+n)!}{k!} \frac{(-\mu t^\nu)^k}{\Gamma(\nu(k+n)+1)}$
$\mu_{N(t)}$	μt	$\frac{\mu t^\nu}{\Gamma(\nu+1)}$
$\sigma^2_{N(t)}$	μt	$\frac{\mu t^\nu}{\Gamma(\nu+1)} \left\{ 1 + \frac{\mu t^\nu}{\Gamma(\nu+1)} \left[\frac{\nu B(\nu,1/2)}{2^{2\nu-1}} - 1 \right] \right\}$,
		$B(\alpha,\beta) = \frac{\Gamma(\alpha)\Gamma(\beta)}{\Gamma(\alpha+\beta)}$
$\mathbf{E}\left[N(t)\right]^k$	$\frac{\partial^k}{\partial s^k} s^k \exp\left[\mu(s-1)t\right]\Big|_{s=0}$	$(-1)^k \frac{\partial^k}{\partial s^k} \sum_{m=0}^{\infty} \frac{\left[\mu t^\nu \left(e^{-s}-1\right)\right]^m}{\Gamma(m\nu+1)}\Big|_{s=0}$

More recently, Mainardi et al. (2004, 2005) provided an approach to fPp based on analysis of the survival probability function $\Theta(t) = P(T > t)$. They have shown that $\Theta(t)$ satisfies the fractional differential equation

$$_0 D_t^{*\nu} \Theta(t) = -\mu\Theta(t), \qquad t \geq 0, \qquad \Theta(0^+) = 1, \tag{3.30}$$

where

$$_0 D_t^{*\nu} f(t) = \begin{cases} \frac{1}{\Gamma(1-\nu)} \int_0^t \frac{f^{(1)}(\tau)}{(t-\tau)^\nu} d\tau, & 0 < \nu < 1; \\[2ex] \frac{d}{dt} f(t), & \nu = 1. \end{cases}$$

is the so-called Caputo derivative. Obviously, for the standard Poisson process with parameter μ, $\Theta(t)$ satisfies the ordinary differential equation

$$\frac{d}{dt}\Theta(t) = -\mu\Theta(t), \qquad t \geq 0, \qquad \Theta(0^+) = 1.$$

Some characteristics of the classical and fractional Poisson processes are compared in Table 1, above.

3. Simulation of fPp interarrival times (T_i). Simulation of the usual Poisson process is very easy and efficient because, given a random variable U, uniformly distributed on $[0, 1]$, the random variable $|\ln U|/\mu$ has the exponential distribution with parameter μ. With the interarrival time for fPp exhibiting a more complicated structure described in Section 2 the issue of an efficient simulation of fPp depends on finding a representation for the Mittag-Leffler function which is more computationally convenient than the series (3.21). Here, the critical observation is that the interarrival times $T_i \stackrel{d}{=} T$, are equidistributed with the random variable,

$$T' = \frac{|\ln U|^{1/\nu}}{\mu^{1/\nu}} S_\nu,$$

where $S_\nu \geq 0$ is a completely asymmetric ν-stable random variable (see, Appendix 1) with the p.d.f. $g_\nu(s)$ possessing the Laplace transform

$$\int_0^\infty g_\nu(s)e^{-\lambda s}\,ds = \exp(-\lambda^\nu). \tag{3.31}$$

The verification of the above statement is straightforward in view of (3.20) and the integral representation for the Mittag-Leffler function implied by (3.31); cf., e.g., Uchaikin and Zolotarev (1999):

$$\begin{aligned}
\mathbf{P}(T' > t) &= \mathbf{P}\left(\frac{|\ln U|^{1/\nu}}{\mu^{1/\nu}}S_\nu > t\right) \\
&= \int_0^1 \mathbf{P}\left(S_\nu > \frac{t\mu^{1/\nu}}{(-\ln(1-u))^{1/\nu}}\right) du \\
&= \int_0^\infty \mathbf{P}\left(S_\nu > \frac{t}{\tau^{1/\nu}}\right)\mu e^{-\mu\tau}\,d\tau \\
&= \int_0^\infty \left(\int_{t/\tau^{1/\nu}}^\infty g_\nu(s)\,ds\right)\mu e^{-\mu\tau}\,d\tau \\
&= \int_0^\infty \int_{t^\nu/s^\nu}^\infty g_\nu(s)\mu e^{-\mu\tau}\,d\tau\,ds \\
&= \int_0^\infty g_\nu(s)e^{-\mu t^\nu/s^\nu}\,ds = E_\nu(-\mu t^\nu) = \mathbf{P}(T > t).
\end{aligned}$$

Utilizing the well known Kanter (-Chambers-Mallows) algorithm (see, e.g., Kanter, 1975) we obtain the following corollary providing an algorithm for simulation of the fPp interarrival times (Figure 3.4):

Corollary 3.2. *Let U_1, U_2, and U_3, be independent, and uniformly distributed in $[0,1]$. Then the fPp interarrival time*

$$T \overset{d}{=} \frac{|\ln U_1|^{1/\nu}}{\mu^{1/\nu}}\frac{\sin(\nu\pi U_2)[\sin((1-\nu)\pi U_2)]^{1/\nu-1}}{[\sin(\pi U_2)]^{1/\nu}|\ln U_3|^{1/\nu-1}}. \tag{3.32}$$

A comparison of sample trajectories for the standard Poisson process and an fPp, with parameter $\nu = 1/2$, can be seen in Figure 3.4.

4. Scaling limit for fractional Poisson distribution. For the standard Poisson process, $N(t) = N_1(t)$, the central limit theorem and infinite divisibility of the Poisson distribution give us immediately the following Gaussian scaling limit of distributions: as $\bar{n} = \mu t \to \infty$,

$$\frac{N(t) - \bar{n}}{\sqrt{\bar{n}}} \overset{d}{\Longrightarrow} N(0,1).$$

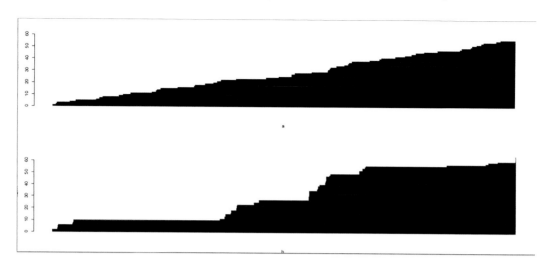

Figure 3.4 Sample trajectories of standard Poisson process (a), and fPp with parameter $\nu = 1/2$ (b).

A more subtle, skew-normal approximation to the Poisson distribution is provided by the following formula: for $n = 0, 1, 2, \ldots,$

$$\mathbf{P}(N(t) \le n) \approx \Phi(z) - \frac{1}{6\sqrt{\bar{n}}}(z^2 - 1)\phi(z),$$

where $z = (n + \frac{1}{2} - \bar{n})/\sqrt{\bar{n}}$, and Φ, and ϕ, are standard normal c.d.f, and p.d.f., respectively. The above formula, used to calculate the probabilities $\mathbf{P}(m < N(t) \le n)$ (including $\mathbf{P}(N(t) = n)$), guarantees, uniformly over n, m, errors not worse than $1/(20\bar{n})$ (as opposed to errors of the order $1/\sqrt{\bar{n}}$ if the skewness correction term is dropped), see, e.g., Pitman (1993), p. 225.

Considering the case of the fPp, $N_\nu(t)$, and introducing the standardized random variable

$$Z_\nu = \frac{N_\nu(t)}{\bar{n}_\nu}, \qquad \text{where} \qquad \bar{n}_\nu = \mathbf{E}N_\nu(t) = \frac{\mu t^\nu}{\Gamma(\nu + 1)},$$

and substituting $u = e^{-\lambda/\bar{n}_\nu}$ in (3.10), we get the Laplace transform

$$\mathbf{E}e^{-\lambda Z_\nu} = E_\nu(\bar{n}_\nu \Gamma(\nu + 1)(e^{-\lambda/\bar{n}_\nu} - 1)), \qquad \lambda > 0,$$

which has, for large $\bar{n}_\nu$ (i.e. large t) the asymptotics

$$\mathbf{E}e^{-\lambda Z_\nu} \sim E_\nu(-\lambda'), \qquad \lambda' = \lambda\Gamma(\nu + 1).$$

Since,

$$E_\nu(-\lambda') = \nu^{-1}\int_0^\infty \exp(-\lambda' x)g_\nu(x^{-1/\nu})x^{-1-1/\nu}dx$$

$$= \int_0^\infty e^{-\lambda z} \left\{ \frac{[\Gamma(\nu+1)]^{1/\nu}}{\nu} g_\nu \left(\left(\frac{z}{\Gamma(\nu+1)} \right)^{-1/\nu} \right) z^{-1-1/\nu} \right\} dz,$$

where $g_\nu(s)$ is the ν-stable p.d.f., see Uchaikin and Zolotarev (1999), formula (6.9.8), the random variable Z_ν has, for $\bar{n}_\nu \to \infty$, a non-degenerate limit distribution with the p.d.f.

$$f_\nu(z) = \left\{ \frac{[\Gamma(\nu+1)]^{1/\nu}}{\nu} g_\nu \left(\left(\frac{z}{\Gamma(\nu+1)} \right)^{-1/\nu} \right) z^{-1-1/\nu} \right\}, \qquad (3.33)$$

with moments

$$\langle Z^k \rangle = \frac{[\Gamma(1+\nu)]^k \Gamma(1+k)}{\Gamma(1+k\nu)},$$

see Uchaikin (1999). Making use of the series expansion for g_ν, we obtain the series expansion

$$f_\nu(z) = \sum_{k=0}^\infty \frac{(-z)^k}{k! \Gamma(1-(k+1)\nu)[\Gamma(\nu+1)]^{k+1}}.$$

Note that

$$f_\nu(0) = \frac{1}{\Gamma(1+\nu)\Gamma(1-\nu)} = \frac{\sin(\nu\pi)}{\nu\pi}.$$

The above family of limiting distributions is plotted below (Figure 3.5).

It is also worth to note, that $\langle Z^0 \rangle = 1$, $\langle Z^1 \rangle = 1$ and $\langle Z^2 \rangle = 2\nu B(\nu, 1+\nu)$, so that the limit relative fluctuations is given by

$$\delta_\nu \equiv \sigma_{N(t)}/\langle N \rangle = \sqrt{2\nu B(\nu, 1+\nu) - 1} = \begin{cases} 1, & \nu = 0, \\ \sqrt{\pi/2} - 1, & \nu = 1/2 \\ 0, & \nu = 1. \end{cases}$$

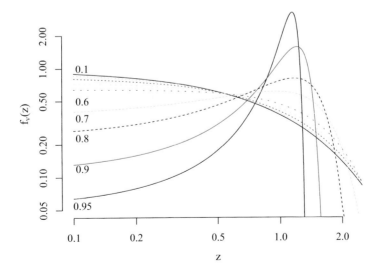

Figure 3.5 Limiting distributions for $\nu = 0.1, ..., 0.6, 0.7, 0.8, 0.9$, and 0.95.

For $\nu = 1/2$, one can obtain an explicit expression for $f_\nu(z)$:

$$f_{1/2}(z) = \frac{2}{\pi} e^{-z^2/\pi}, \ z \geq 0.$$

3.6 PROBLEMS AND EXERCISES

3.4.1. Calculate the n-dimensional distributions of the standard Poisson process.

3.4.2. Show that the joint probability of the waiting times of the Poisson process

$$\mathbf{P}(W_1 < t_1, \ldots, W_n < t_n | N(t) = n)$$

has the uniform density $n!/t^n$, over the simplex $0 < t_1 < \cdots < t_n < t$ in $\mathbf{R}^n_+$.

3.4.3. Prove that the characteristic function

$$\varphi_{W_n}(u) = (1 - i\mu^{-1}u)^{-n}$$

and that $\mathbf{E}W_n = \mu^{-1}n$, and Var $W_n = \mu^{-2}n$.

CHAPTER 4

Lévy Processes and the Lévy-Khinchine Formula: Basic Facts

4.1 PROCESSES WITH STATIONARY AND INDEPENDENT INCREMENTS

In this chapter, we provide a brief, *ab ovo*, review of general stochastically continuous processes with stationary and independent increments also called Lévy processes. Two basic examples of such processes, the Brownian motion and the Poisson process, residing on the opposite ends of the spectrum of Lévy processes, are discussed in the preceding Chapters.

Definition 4.1. Process $X(t), t \geq 0$ is said to have *stationary (homogeneous) increments* if, for any $t_1, t_2 \geq 0, h > 0$, the distributions of the increments,

$$X(t_1 + h) - X(t_1), \quad \text{and} \quad X(t_2 + h) - X(t_2),$$

are identical.

Definition 4.2. Process $X(t), t \geq 0$ is said to have *independent increments* if, for any $0 < t_1 < \cdots < t_n$, the increments

$$X(t_1) - X(0), X(t_2) - X(t_1), \ldots, X(t_n) - X(t_{n-1})$$

are independent random variables.

The processes that satisfy both conditions are called, unsurprisingly, *processes with stationary and independent increments* (SII processes in brief). We will also assume, in general, that $X(0) = 0$.

The main observation about the SII processes is that their one-dimensional distributions completely determine the structure of their d-dimensional distributions for any d.

Let

$$F_{t_1,\dots,t_d}(x_1,\dots x_d) = \mathbf{P}(X(t_i) \le x_i, \; i=1,\dots,d)$$

be the d-dimensional cdf and

$$\varphi_{t_1,\dots,t_d}(u_1,\dots u_d) = \int_{-\infty}^{\infty} e^{i(u_1 x_1 + \cdots + u_d x_d)} dF_{t_1,\dots,t_d}(x_1,\dots x_d),$$

the d-dimensional characteristic function that uniquely determines F itself. Then, indeed,

$$\varphi_{t_1,\dots,t_d}(u_1,\dots u_d) = \mathbf{E}\exp\left(i\sum_{j=1}^{d} u_j X(t_j)\right)$$

$$= \mathbf{E}\exp\Big(i(u_1 + u_2 + \cdots + u_d)\big(X(t_1) - X(0)\big)$$

$$+ i(u_2 + \cdots + u_d)\big(X(t_2) - X(t_1)\big)$$

$$+ \cdots + iu_d\big(X(t_d) - X(t_{d-1})\big)\Big)$$

$$= \varphi_{t_1}(u_1 + \cdots + u_d) \cdot \varphi_{t_2 - t_1}(u_2 + \cdots + u_d) \cdot \dots \cdot \varphi_{t_d - t_{d-1}}(u_d).$$

4.2 FROM POISSON PROCESSES TO LÉVY PROCESSES

Let us go back to an elementary one-dimensional example of the Poisson process, where a particle is located on a 1-D lattice with unit spacing, starts at the origin, waits a random exponential time and then moves one unit to the right. Then, the step is independently repeated. If we denote by $N(t)$, the position of the particle at time $t > 0$, then, as we have seen in Chapter 3, this random variable has the standard Poisson distribution, i.e.,

$$\mathbf{P}(N(t) = k) = e^{-t} \cdot t^k/k!, \qquad k = 0, 1, 2, \dots.$$

This distribution can be described in terms of its characteristic function (Fourier transform) $\varphi(u)$ as follows

$$\varphi(u) = \mathbf{E}e^{iuN(t)} = e^{-t}\sum_{k=0}^{\infty}(e^{iu})^k \cdot \frac{t^k}{k!} = \exp[t(e^{iu} - 1)] \tag{4.1}$$

If the jumps are of size (amplitude) a, i.e.,

$$\mathbf{P}(N(t) = ka) = e^{-t} \cdot t^k/k!, \qquad k = 0, 1, 2, \dots.$$

then the analogous calculation gives the characteristic function

$$\varphi(u) = \mathbf{E}e^{iuX(t)} = e^{-t}\sum_{k=0}^{\infty}(e^{iu})^{ak} \cdot \frac{t^k}{k!} = \exp[t(e^{iua} - 1)] \tag{4.2}$$

Now consider a more complex model that is a composition of n independent simple Poisson processes, each with jump sizes $a_1, \ldots, a_n$, and means $\mu_1, \ldots, \mu_n$, respectively. For the resulting stochastic process $X(t)$, the characteristic function

$$\varphi(u) = \mathbf{E}e^{iuX(t)} = \prod_{j=1}^{n} \sum_{k=0}^{\infty} e^{-t\mu_j}(e^{iu})^{a_j k} \cdot \frac{(t\mu_j)^k}{k!} = \exp\left(t\sum_{j=1}^{n}(e^{iua_j} - 1)\mu_j\right).$$

(4.3)

If jumps sizes are continuously distributed, say with intensity $L(da)$, then the natural infinitesimal limiting procedure for the corresponding *Lévy process* leads formally to the following representation of its Fourier transform

$$\varphi(u) = \mathbf{E}e^{iuX(t)} = \exp\left(t\int_{-\infty}^{\infty}(e^{iua} - 1)\,L(da)\right)$$

assuming that the time increments are stationary and independent in disjoint time intervals. Note that to guarantee the existence of the above integral, the measure L has to be finite, in which case we have

$$\left|\int_{-\infty}^{\infty}(e^{iua} - 1)\,L(da)\right| \leq \int_{-\infty}^{\infty}|e^{iua} - 1|\,L(da) \leq 2L(\mathbf{R}).$$

If measure L has a density with respect to the Lebesgue measure, that is, $L(da) = \lambda(a)\,da$, then the above condition requires $\lambda(a)$ to be integrable,

$$\int_{-\infty}^{\infty} \lambda(a)\,da < \infty.$$

To permit a wider class of processes with stationary and independent increments to be included and, in particular, to expand the class of self-similar (Gaussian) processes discussed so far (Section 4.4), it is necessary to permit the density $\lambda(a)$ to have a non-integrable singularity in the neighborhood of 0. To accomplish this goal, we will rewrite the above expression for φ in the form

$$\varphi(u) = \mathbf{E}e^{iuX(t)} = \exp\left(t\int_{-\infty}^{\infty}\left(e^{iua} - 1 - iuaI_{[-1,1]}(a)\right)L(da)\right)$$

(4.4)

which gives the so-called Lévy-Khinchine formula for the characteristic function of a general stochastically continuous process with stationary and independent increments (Feller (1966), Bertoin (1996)). Now, the Lévy measure has to satisfy the following weaker integrability condition:

$$\int_{-\infty}^{\infty}(1 \wedge a^2)\,L(da) < \infty.$$

(4.5)

In the case of a symmetrically distributed Lévy process, this characteristic function is real-valued and of a simpler form,

$$\varphi(u) = \exp\left(t\int_{-\infty}^{\infty}\left(\cos ua - 1\right)L(da)\right).$$

(4.6)

It will be convenient to define the characteristic exponent $\Psi(u)$ of the Lévy process by the formula

$$\Psi(u) = \int_{-\infty}^{\infty} \left(e^{iua} - 1 - iuaI_{[-1,1]}(a) \right) L(da). \tag{4.7}$$

Then its characteristic function $\varphi(u) = \exp(t\Psi(u))$.

4.3   INFINITESIMAL GENERATORS OF LÉVY PROCESSES

Lévy processes are Markov processes with the associated Markov semi-group (i.e. $P_{t+s} = P_t P_s$, $t, s > 0$,) of convolution operators P_t acting on a bounded function $f(x)$ via the formula

$$P_t f(x) = Ef(x - X(t)) = \int_R f(x-y) \, P(X(t) \in dy) = \int_R f(x-y) f_{X(t)}(y) \, dy. \tag{4.8}$$

Indeed,

$$
\begin{aligned}
P_{t+s} f(x) &= \int_R f(x-y) f_{X(t+s)}(y) \, dy = \int_R f(x-y)[f_{X(t)} * f_{X(s)}](y) \, dy \\
&= \int_R f(x-y) \left[\int_R f_{X(t)}(z) \cdot f_{X(s)}(y-z) \, dz \right] dy \\
&= \int_R \int_R f(x - w - z) \cdot f_{X(t)}(z) \cdot f_{X(s)}(w) \, dz \, dw = P_t P_s f(x).
\end{aligned}
$$

Also not that for $f(x) = \delta(x)$

$$P_t \delta(x) = \int_R \delta(x-y) f_{X(t)}(y) \, dy = f_{X(t)}(x)$$

gives the evolution of the 1-D densities of the process $X(t)$.

The *infinitesimal generator* $\mathcal{A}$ of such a semi-group is defined by the formula

$$\mathcal{A} = \lim_{h \to 0} \frac{P_h - P_0}{h} \tag{4.9}$$

and the family of functions (densities) $v(t, x) = P_t f(x)$ satisfies clearly the (generalized) Fokker–Planck evolution equation

$$\frac{\partial v}{\partial t} = \mathcal{A}v, \tag{4.10}$$

because

$$\lim_{h \to 0} \frac{P_{t+h} - P_t}{h} = \lim_{h \to 0} \frac{P_h - P_0}{h} P_t = \mathcal{A}P_t. \tag{4.11}$$

In the case of the usual Brownian motion, the infinitesimal operator $\mathcal{A}$ is just the classical Laplacian Δ. In the case of general Lévy processes, we have the identity

$$\mathcal{F}(\mathcal{A}f)(u) = \Psi(-u)\mathcal{F}f(u) \tag{4.12}$$

where $\mathcal{F}$ stands for the Fourier transform, because

$$\mathcal{F}(P_t f)(u) = E\left(\int_R e^{-iux} f(x - X(t))\, dx\right) = E\left(\int_R e^{-iu(y + X(t))} f(y)\, dy\right)$$
$$= E e^{-iuX(t)} \int_R e^{-iuy} f(y)\, dy = \exp(t\Psi(-u))\mathcal{F}f(u),$$

so that

$$(P_t f)(x) = \mathcal{F}^{-1}[\exp(t\Psi(-u))\mathcal{F}f(u)](x),$$

Inverting the above Fourier transform, one gets the following representation for the infinitesimal operator of the Lévy process

$$\mathcal{A}f(x) = \mathcal{F}^{-1} \lim_{h \to 0} \left[\frac{e^{h\Psi(-u)} - e^{0\Psi(-u)}}{h} \mathcal{F}f(u)\right](x)$$
$$= \mathcal{F}^{-1}[\Psi(-u)\mathcal{F}f(u)](x)$$
$$= \int_R \left(f(x + y) - f(x) - f'(x) \cdot y \cdot I_{[-1,1]}(y)\right) L(dy) \qquad (4.13)$$

4.4 SELF-SIMILAR LÉVY PROCESSES

Brownian motion enjoyed the self-similarity property

$$B_{ct} =_d c^{1/2} B_t,$$

or, in terms of its characteristic function $\varphi_t(u) = e^{-tu^2/2}$,

$$\varphi_t(u) = \varphi_1(t^{1/2}u), \qquad \forall t, u.$$

Therefore, the question arises whether one can find Lévy processes which are self-similar, perhaps with the parameter $\alpha \neq 2$, i.e., satisfying the condition

$$\varphi_t(u) = \varphi_1(t^{1/\alpha}u), \qquad \alpha \neq 2? \qquad (4.14)$$

Focussing on the symmetric case with the Lévy measure admitting a density, $L(da) = \lambda(a)\, da$, the self-similarity condition (5.1) can be rewritten in the form,

$$\exp\left(t \int_0^\infty (\cos ua - 1)\lambda(a)\, da\right) = \exp\left(\int_0^\infty (\cos(t^{1/\alpha}ua) - 1)\lambda(a)\, da\right).$$

After a substitution, $a = t^{-1/\alpha}z$ on the right-hand side, the condition becomes a functional equation

$$\lambda(a) = \lambda(t^{-1/\alpha}a)t^{-1-1/\alpha}, \qquad \forall t, a.$$

Taking $t = a^\alpha$, we thus get the functional form of the density of the self-similar Lévy measure,

$$\lambda(a) = \lambda(1)a^{-\alpha-1}. \qquad (4.15)$$

where $\lambda(1)$ can be an arbitrary positive constant.

The parameter α cannot be arbitrary. The Lévy measure integrability condition of Chapter 4,

$$\int_0^\infty (1 \wedge a^2) \frac{da}{a^{\alpha+1}} < \infty$$

forces the condition $0 < \alpha < 2$.

Such processes X_t are called symmetric α-stable processes (αS) or α-stable motions. Their characteristic function is

$$\varphi_t(u) = E \exp(iuX_t) = \exp\left(t \int_{-\infty}^\infty (\cos ux - 1) \frac{dx}{|x|^{\alpha+1}} \right) = e^{-C_\alpha t |u|^\alpha}, \qquad (4.16)$$

where $C = \int_{-\infty}^\infty (1 - \cos z) z^{-\alpha-1} \, dz$ is a constant. The stability property, or the self-similarity property, can now be rephrased in the distributional terms as follows:

$$X_{ct} =_d c^{1/\alpha} X_t. \qquad (4.17)$$

Examples. (a) Poisson Processes and generalized Poisson processes are Levy processes with obvious discrete Lévy measures, but they are not self-similar.

(b) For $\alpha = 1$, the α-stable characteristic function $e^{-|u|}$ is easily recognizable as that of the Cauchy distribution with pdf $f_{X(1)}(x) = 1/(\pi(1+x^2))$. Other symmetric α-stable pdfs cannot be calculated in terms of elementary functions.

(c) The Brownian motion is, of course, two-stable in terms of the definition of self-similarity, but cannot be squeezed rigorously into the above scheme. However, calculating informally with the Dirac-delta measure, one can observe that taking

$$\lambda_\epsilon(a) = \frac{\delta(a - \epsilon)}{a^2},$$

we have,

$$\lim_{\epsilon \to 0} \int_0^\infty (\cos ua - 1) \lambda_\epsilon(a) \, da = \lim_{\epsilon \to 0} \frac{\cos u\epsilon - 1}{\epsilon^2} = -\frac{u^2}{2}.$$

4.5   PROPERTIES OF α-STABLE MOTIONS

Since

$$\int_{0 < |a| \le 1} dL(a) = \int_{0 < |a| \le 1} \frac{da}{|a|^{\alpha+1}} = \infty$$

the α-stable motion process is not a pure jump process, i.e., its sample paths are not piecewise constant a.s. A more precise description of their structure will be discussed later on.

The one-dimensional distributions no longer have finite second moments. More precisely,

Theorem 4.1. *The tail probabilities of $X = X(1)$ decay in a power-type fashion:*

$$\lim_{x \to \infty} x^\alpha \mathbf{P}(|X| > x) = C < \infty \qquad (4.18)$$

Proof. Indeed, introducing a new auxiliary Lévy measure

$$\Lambda_n(\{x : |x| > z\}) := n\mathbf{P}\left(\frac{|X|}{n^{1/\alpha}} > z\right), \quad z \geq 0,$$

we have, $\forall u$,

$$\int (e^{iux} - 1)\Lambda_n(dx) = n\left(e^{-|u|^\alpha/n} - 1\right) \longrightarrow -|u|^\alpha = C\int (e^{iux} - 1)\frac{dx}{|x|^{\alpha+1}},$$

as $n \to \infty$, because $n(e^{z/n} - 1) \to z$. Consequently, as $n \to \infty$,

$$n\mathbf{P}\left(\frac{|X|}{n^{1/\alpha}} > z\right) \longrightarrow 2\int_z^\infty \frac{dx}{x^{\alpha+1}} = 2z^{-\alpha},$$

which yields (5.5.) □

Corollary 4.1. *(a) Moments of X of order equal to or greater than α are infinite. Indeed,*

$$\int_0^\infty x^\alpha f_\alpha(x)\,dx = -x^\alpha(1 - F_\alpha(x))\,\Big|_0^\infty + \alpha\int_0^\infty x^{\alpha-1}(1 - F_\alpha(x))\,dx = \infty$$

because $(1 - F_\alpha(x)) \sim x^{-\alpha}$.
   (b) Moments of X of order $0 < \beta < \alpha$ are finite. Indeed,

$$\int_0^\infty x^\beta f_\alpha(x)\,dx = -x^\beta(1 - F_\alpha(x))\,\Big|_0^\infty + \beta\int_0^\infty x^{\beta-1}(1 - F_\alpha(x))\,dx < \infty$$

Trajectories of α-stable motions have a.s. nontrivial fractional (Hausdorff)dimension.

Remark 4.1 (Chambers-Mallows-Stuck simulation algorithm). Since, with a few exceptions, it is impossible to explicitly calculate the CDF of the α-stable distribution, and its inverse, the usual standard method of generating random variables is not available. In this context, there were many efforts to produce such an algorithm and one of them, due to Chambers, Mallows and Stuck[1] depends on the observation that, if $0 < \alpha \leq 2$, U is a random variable uniformly distributed on $[-\pi/2, \pi/2]$, and V is an independent exponential random variable with mean 1, then

$$S_\alpha = \frac{\sin(\alpha U)}{[\cos(U)]^{1/\alpha}} \cdot \left[\frac{\cos(U - \alpha U)}{V}\right]^{\frac{1-\alpha}{\alpha}},$$

is a standard α-stable random variable.

---

[1] J.M. Chambers, C.L. Mallows and B.W. Stuck, A method for simulating stable random variables, *J. Amer. STat. Assoc.* **71**(1976), 340–344.

4.6 INFINITESIMAL GENERATORS OF α-STABLE MOTIONS

In the special case of the standard α-stable Lévy process to be discussed in Chapter 5, i.e., when $\Psi(u) = -|u|^\alpha$, the infinitesimal generator is of the form

$$\mathcal{A}_\alpha f(x) = \mathcal{F}^{-1}(-|u|^\alpha)(\mathcal{F}f)(u) \tag{4.19}$$

so it can be identified as the derivative D^α, of order α. In the case $\alpha = 2$, it gives just the second derivative. Therefore, the corresponding Fokker–Planck diffusion equation for the densities of the α-stable Lévy process is of the form

$$\frac{\partial}{\partial t} f_{X(t)}(x) = D^\alpha f_{X(t)}(x)$$

in complete analogy to the heat equation

$$\frac{\partial}{\partial t} f_{X(t)}(x) = \frac{\partial^2}{\partial^2 x} f_{X(t)}(x)$$

describing the evolution of 1-D densities of the Brownian motion ($\alpha = 2$).

It is also worthwhile to observe that the formula

$$\mathcal{A}_\alpha f(x) = \int_R (f(x+y) - f(x)) \frac{dy}{|y|^{\alpha+1}} \tag{4.20}$$

gives an explicit expression for the infinitesimal generator of the α-stable process. Indeed, since in the Fourier multiplier operator of the form

$$\mathcal{F}(\mathcal{A}_\alpha f)(u) = \int_R e^{-iux} \left(\int_R (f(x+y) - f(x)) \frac{dy}{|y|^{\alpha+1}} \right) dx$$

$$= \int_R \left(\int_R e^{-iux} f(x+y)\, dx - \int_R e^{-iux} f(x)\, dx \right) \frac{dy}{|y|^{\alpha+1}}$$

$$\int_R (\mathcal{F}f)(u) \cdot (e^{iuy} - 1) \frac{dy}{|y|^{\alpha+1}} = -C(|u|)^\alpha \cdot (\mathcal{F}f)(u)$$

Remark 4.2. In the multidimensional case, the infinitesimal generator of the Brownian motion (i.e., two-stable motion) was the usual Laplacian (second derivative in the one-dimensional case) Δ, for which the Fourier multiplier was just $-|u|^2$ and for the analogous α-stable diffusion the generator is the fractional power of the Laplacian.

The above exposition only gives a sketch of the fractional Laplacian machinery. Some mathematical details have been omitted not to cloud the basic formal structure. Those details can be found in E.M., Feller(1966), Bertoin(1996), and Saichev and Woyczynski (1997).

4.7   PROBLEMS AND EXERCISES

4.7.1. Verify the last line of the expression (4.13) for the infinitesimal generator of the Lévy process.

4.7.2. Continuous Time Random Walks (CTRW)

4.7.3. Provide simulations of Lévy stable processes for several choices of α using both the Chambers-Mallows algorithm and the mixed Poisson approximations. Compare the results.

CHAPTER 5

General Processes with Independent Increments

5.1 NONSTATIONARY PROCESSES WITH INDEPENDENT INCREMENTS

If we remove the stationarity-of-increments restriction in the model discussed above, then the knowledge of 1-D distributions is no longer sufficient to determine all the finite-dimensional distributions. However, the knowledge of the 2-D distributions of the increments suffices. Indeed, with

$$F_{t_1,\ldots,t_d}(x_1,\ldots,x_d) = \mathbf{P}(X(t_i) \le x_i, i = 1, 2, \ldots, d)$$

denoting, as before, the d-dimensional CDFs, and

$$\varphi_{t_1,\ldots,t_d}(u_1,\ldots,u_d) = \mathbf{E}\exp\left\{ i\sum_{j=1}^d u_j X(t_j) \right\}$$

$$= \int_{-\infty}^{\infty} \cdots \int_{-\infty}^{\infty} e^{i(u_1 x_1 + \cdots + u_d x_d)} dF_{t_1,\ldots,t_d}(x_1,\ldots,x_d)$$

standing for the d-dimensional characteristics function of the random vector $X(t_1),\ldots,X(t_d))$, denoting

$$\psi_t(u) = \mathbf{E}e^{iuX(t)},$$

and

$$\psi_{t_1,t_2}(u) = \mathbf{E}e^{iu(X(t_2 - X(t_1)))},$$

we see that

$$\varphi_{t_1,\ldots,t_d}(u_1,\ldots,u_d) = \mathbf{E}\exp\Big\{ i(u_1 + u_2 + \cdots + u_d)X(t_1)$$
$$+ i(u_2 + \cdots + u_d)(X(t_2) - X(t_1))$$
$$+ i(u_3 + \cdots + u_d)(X(t_3) - X(t_2))$$
$$\cdots\cdots\cdots\cdots\cdots$$
$$+ iu_d(X(t_d) - X(t_{d-1}))\Big\}$$
$$= \psi_{t_1}(u_1 + \cdots + u_d) \cdot \psi_{t_1,t_2}(u_2 + \cdots + u_d) \cdot \ldots \cdot \psi_{t_{d-1},t_d}(u_d).$$

DOI: 10.1201/9781003216759-5

Example 5.1 (Nonhomogeneous Poisson process). *The standard Poisson process model discussed above can be extended by permitting the intensity μ, see Section 3.1., to vary in time. In this case, the 1-D distributions are*

$$\mathbf{P}(X(t) = k) = e^{\mu(t)} \frac{\mu^k(t)}{k!}, \qquad t \geq 0, \quad k = 0, 1, 2, \ldots, \tag{5.1}$$

where $\mu(t)$ is a nonnegative nondecreasing function on the positive half-line, with the distributions of the increments over the intervals $[t_1, t_2]$

$$\mathbf{P}(X(t_2) - X(t_1) = k) = e^{\mu(t_2) - \mu(t_1)} \frac{(\mu(t_2) - \mu(t_1))^k}{k!}. \tag{5.2}$$

In this case, the 1-D characteristic function is

$$\varphi_t(u) = \exp[\mu(t)(e^{iu} - 1)], \tag{5.3}$$

the increments $X(t_2) - X(t_1)$ have the characteristic function

$$\psi_{t_1,t_2}(u) = \exp\big[(\mu(t_2) - \mu(t_2))(e^{iu} - 1)\big] \tag{5.4}$$

Example 5.2 (A generalized Brownian motion). *Let's consider a general Gaussian process $X(t)$ with independent increments. Assume $X(0) = 0$, and that the mean and variance are*

$$\mathbf{E}X(t) = a(t), \qquad \text{Var } (X(t) = b(t).$$

For the increments we have, with $t_1 < t_2$,

$$\mathbf{E}(X(t_2) - X(t_1)) = a(t), \qquad \text{Var}(X(t_2) - X(t_1)) = b(t_2) - b(t_1),$$

because, in view of the independence of increments, $\text{Var}(X(t_2)) = \text{Var}(X(t_1)) + \text{Var}(X(t_2) - X(t_1))$.

The CDFs of $X(t)$ and is increments are as follows:

$$\mathbf{P}(X(t) \leq x) = \frac{1}{\sqrt{2\pi b(t)}} \int_{-\infty}^{x} \exp\left\{-\frac{(z - a(t))^2}{2b(t)}\right\} dz,$$

and

$$\mathbf{P}(X(t_2) - X(t_1) \leq x) = \frac{1}{\sqrt{2\pi(b(t_2) - b(t_1))}} \int_{-\infty}^{x} \exp\left\{-\frac{(z - [a(t_2) - a(t_1)])^2}{2(b(t_2) - b(t_1))}\right\} dz,$$

so that one- and two-dimensional characteristic functions determining all the finite-dimensional distributions are as follows:

$$\psi_t(u) = \exp\left\{iua(t) - u^2 b(t)/2\right\},$$

$$\psi_{t_1,t_2}(u) = \exp\left\{iu[a(t_2) - a(t_1)] - u^2[b(t_2) - b(t_1)]/2\right\}.$$

Example 5.3 (Processes with random jumps at fixed times). *Let $\xi_1, \xi_2, \ldots$ be a sequence of independent random variables such that the random series $\sum_i \xi_i$ converges unconditionally, and let $0 \leq t_1 < t_2 < \cdots < t_n$. Define a stochastic process $X(t)$ by the formula*

$$X(t) = \sum_{\{k:t_k < t\}} \xi_k.$$

The process $X(t)$ has independent increments and $X(0) = 0$ so that to obtain its finite-dimensional distributions it is sufficient to calculate the distributions of its increments $X(t) - X(s)$, $s < t.$,
   If

$$\eta_n = X(t) = \sum_{\{k \leq n:t_s \leq t_k < t\}} \xi_k.$$

then, as $n \to \infty$,

$$\mathbf{P}\left(\eta_n \to X(t) - X(s)\right) = 1$$

and

$$\exp\{iu\eta_n\} \to \exp\{iu(X(t) - X(s))\}$$

in probability, so that, by the Lebesgue Bounded Convergence Theorem,

$$\mathbf{E}\exp\{iu\eta_n\} \to \psi_{s,t}(u),$$

and, we finally get, with $\varphi_k(u) = \mathbf{E}\exp(iu\xi_k)$, that

$$\psi_{s,t}(u) = \lim_{n \to \infty} \prod_{\{k \leq n:t_s \leq t_k < t\}} \varphi_k(u).$$

For a stock market example of the pure jump process at fixed times mixed with the continuous Brownian motion-type continuous process, see Figure 5.1.

5.2 STOCHASTIC CONTINUITY AND JUMP PROCESSES

Definition 5.1. Process $X(t)$ is said to be *stochastically continuous* if for each t_0, and each sequence $t_n \to t_0$, we have $X(t_n) \xrightarrow{\mathbf{P}} X(t_0)$, that is, for each $\epsilon > 0$,

$$\lim_{n \to \infty} \mathbf{P}(|X(t_0) - X(t_0)| > \epsilon) = 0.$$

For processes with independent increments, the assumption of stochastic continuity is equivalent to the assumption that the one-dimensional characteristic function $\varphi_t(u) = \mathbf{E}\exp(iuX(t))$ is a continuous function of the time variable t.

The importance of stochastic continuity assumption is that it guarantees that the sample paths have only simple jump discontinuities but no oscillatory-time discontinuities.

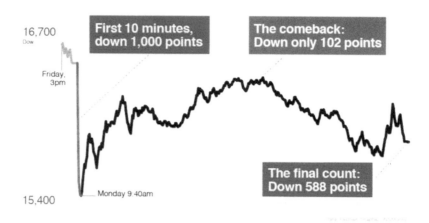

Figure 5.1 Stock market prices typically have a jump between the closing and the opening of the market on the next trading day. Therefore, the process can be viewed as a mixture of a Brownian motion and pure jump process with random jumps at fixed times.

Theorem 5.1. *If $X(t)$, $t \in [a,b]$ is a separable[1], stochastically continuous process with independent increments then, with probability 1, $X(t)$ has no discontinuities of the second kind, that is, its sample path have left and right limits at each point $t \in [a,b]$.*

Proof. □

Remark 5.1 (Infinite divisibility). Let us also note that any stochastically continuous process with independent increments has infinitely divisible distributions[2] of its increments. This is obvious in the case of stationary increments. Indeed, since

$$X(t+s) = (X(t+s) - X(t)) + (X(t) - X(0)),$$

we have the functional equation

$$\varphi_{t+s}(u) = \varphi_t(u)\varphi_s(u)$$

satisfied for every t, s, and u. The solution is then necessarily exponential,

$$\varphi_t(u) = [\varphi_i(u)]^t = e^{t\psi(u)} = e^{(t/n)\psi(u)} \cdot \ldots \cdot e^{(t/n)\psi(u)}$$

for any n.

Now let us consider a pure jump process $X(t), t \geq 0$, generated as follows: Let $W_1, W_2, \ldots,$ be the jump times of a Poisson process $Y(t)$ with parameter λ, that

---

[1] Separability of a process means .....

[2] A random variable ξ is said to have an infinitely divisible distribution if, for each $n = 1, 2, \ldots,$ $\mathcal{L}(\xi) = \mathcal{L}(\xi_1 + \cdots + \xi_n)$ for some i.i.d random variables $\xi_1, \ldots, \xi_n$.

is $W_n = T_1 + \cdots + T_n$, where $\{T_n\}$ is a sequence of independent random variables (inter-jump times) with common exponential distribution, and let $\xi_1, xi_2, \ldots$, be a sequence of independent random variables with common CDF $F(x)$, and independent of $Y(t)$.

Proposition 5.1. *The pure jump process $X(t)$ defined by assigning jumps ξ_n at times W_n, for $n = 1, 2, \ldots$, has stationary and independent increments and its one-dimensional characteristic function is of the form*

$$\varphi_{X(t)}(u) = \mathbf{E} \exp[iuX(t)] = \exp\left[\lambda t \int_{-\infty}^{\infty} (e^{iux} - 1)\, dF(x)\right]. \qquad (5.5)$$

Proof. The stationarity and independence of increments follow directly from the construction. The characteristics function can be calculated as follows: first, conditionally,

$$\mathbf{E}\left(\exp[iuX(t)] \,\Big|\, Y(t) = n\right) = \mathbf{E}\exp[iu(\xi_1 + \cdots + \xi_n)] = (\varphi_\xi(u))^n,$$

where $\varphi_\xi(u) = \int e^{iux}\, dF(x)$ is the common characteristic function of $\xi_1, \xi_2, \ldots$. Using the definition of conditional expectation, we then obtain

$$\mathbf{E}\exp[iuX(t)] = \sum_{n=0}^{\infty} (\varphi_\xi(u))^n \frac{(\lambda t)^n e^{-\lambda t}}{n!} = e^{\lambda t(\varphi_\xi(u) - 1)}. \qquad \square$$

Therefore, in a sense, one can view a pure jump process as a (continuous) mixture of Poisson processes with different (and random) jump sizes.

Remark 5.2. Defining the finite positive measure on $\mathbf{R}$ via the formula $\mu(dx) = \lambda dF(x)$, we can write the characteristic function of a pure jump process in the form

$$\varphi_{X(t)}(u) = \mathbf{E}\exp[iuX(t)] = \exp\left[t \int_{-\infty}^{\infty} (e^{iux} - 1)\, \mu(dx)\right]. \qquad (5.6)$$

Such a process has sample paths that are constant except for random jumps at a discrete sequence of random times. This is the simplest form of the so-called Lévy-Khinchine formula for general processes with stationary and independent increments.

5.3   ANALYSIS OF JUMP STRUCTURE

Consider disjoint subsets $B_1, \ldots, B_n$, of $\mathbf{R}$ such that

$$\bigcup_{k=1}^{n} B_k = \mathbf{R} \setminus \{0\}.$$

We shall say that $X(t)$ has a jump of size B at time t if $X(t) - X(t-) \in B$. For each $k = 1, \ldots, n$, define a new process

$$X(B_k, t) := \sum_{\tau \leq t} I_{B_k}\left(X(\tau) - X(\tau-)\right) \cdot \left(X(\tau) - X(\tau-)\right),$$

which is simply the sum of all jumps of $X(t)$ of size B_k that occurred up to time t.

Theorem 5.2. *The processes $X(B_1, t), \ldots, X(B_n, t)$, $t \geq 0$, are independent of each other, and each of them has stationary and independent increments and the characteristic function*

$$\varphi_{X(B_k,t)}(u) = \exp\left[t \int_{B_k} [e^{iux} - 1] \mu(dx)\right], \qquad k = 1, \ldots, n.$$

Proof. Let us start from the other end and construct independent processes $\tilde{X}^{(1)}(t), \ldots, \tilde{X}^{(n)}(t)$, with independent increments, and such that

$$\varphi_{X^{(k)}(t)}(u) = \exp\left[t \int_{B_k} [e^{iux} - 1] \mu(dx)\right], \qquad k = 1, \ldots, n.$$

Each of these processes corresponds to the process defined in (5.5) with $\lambda = \mu(B_k)$, and

$$F_k(x) = \frac{\mu((-\infty, x] \cap B_k)}{\mu(B_k)},$$

so that $\tilde{X}^{(k)}(t)$ has only jumps of size B_k.

Now, define

$$\tilde{X}(t) = \sum_{k=1}^{n} \tilde{X}^{(k)}(t), \qquad t \geq 0,$$

which has stationary and independent increments, has the same distributions as $X(t) < t \geq 0$, and satisfies the equality

$$\tilde{X}^{(k)}(t) := \sum_{\tau \leq t} I_{B_k}\left(\tilde{X}(\tau) - \tilde{X}(\tau-)\right) \cdot \left(X(\tau) - X(\tau-)\right),$$

Since the processes $X(B_k, t), t\ ge0, k = 1, \ldots, n$, have been obtained from the process $X(t)$ by application of the same function which yielded $\tilde{X}^{(k)}(t)$ from $\tilde{X}(t)$, and since $X(t)$ and $\tilde{X}(t)$ have the same distribution, we conclude that $X(B_k, t), t \geq 0$, and $\tilde{X}^{(k)}(t), t \geq 0$, also necessarily have the same distribution. □

5.4 RANDOM MEASURES AND RANDOM INTEGRALS ASSOCIATED WITH JUMP PROCESSES

5.4.1 Random Measures and Random Integrals

Consider a set $A \subset \mathbf{R} - [-\epsilon, \epsilon], \epsilon > 0$, and a stochastically continuous SII process $X(t)$, and define

$$N(A, , t) = \sum_{s \leq t} I_A(X(s) - X(s-)),$$

which, plainly speaking, is the number of jumps of $X(t)$ of size A up to time t. Since $X(t)$ has no discontinuities of the second kind, $N(A, t)$ is an integer-valued random variable. We have excluded the ϵ-neighborhood of the origin from the set A because the number of jumps of size $> \epsilon$, for $s \leq t$, is always finite; this is not always the case for small jump sizes in the neighborhood of 0.

One can show that:

(a) For a fixed set $A \subset \mathbf{R} \setminus [-\epsilon, \epsilon]$, $N(A, t), t \geq 0$, is a process with stationary and independent increments.

(b) For a fixed t, $N(A, t)$ is a *random measure* on the σ-field $\mathcal{B}_\epsilon = (\mathbf{R} \setminus [-\epsilon, \epsilon]) \cap \mathcal{B}$, which means that, for pairwise disjoint sets $A_1, A_2, \cdots \in \mathcal{B}_\epsilon$,

$$N \left(\bigcup_{n=1}^{\infty} A_n, t \right) = \sum_{n=1}^{\infty} N(A_n, t),$$

where the series converges in probability. The random measure $N(A, t)$ is *independently scattered*, that is, if $A \cap B = \emptyset$ then the random variables $N(A, t)$ and $N(B, t)$ are independent.

Now, for any bounded and measurable function $h : \mathbf{R} \setminus [-\epsilon, \epsilon] \mapsto \mathbf{R}$, we can define a *random integral* of h with respect to measure N,

$$\int h(x) \, N(dx, t),$$

by first constructing it for step functions h, and the proving the if a sequence (h_n) of step functions converges to h point-wise then, for each bounded A,

$$\int_A h_n(x) \, N(dx, t) \longrightarrow \int_A h(x) \, N(dx, t),$$

in probability, as $n \to \infty$.

The key observations are that:

(c) Process X has a random integral representation in terms of the random measure N:

$$X(A, t) = \int_A x \, N(dx, t)$$

(d) If one denotes

$$\pi(A, t) = \mathbf{E} N(A, t),$$

then, for a fixed A, $N(A, t)$ is a Poisson process with paramater $\pi(A, t)$; in the stationary case, $\pi(A, t) = t \pi(A, 1)$.

5.5   STRUCTURE OF GENERAL I.I. PROCESSES

Let $X(t), t \geq 0$, be an II process (as always, stochastically continuous) and consider a sequence $\epsilon_0 = 1 > \epsilon_n \to 0$ so that, for the intervals $\Delta_n- = (\epsilon_{n+1}, \epsilon_n], n = 0, 1, 2, \ldots$, we have

$$\bigcup_{n=0}^{\infty} \Delta_n = (0, 1].$$

Now, consider the process

$$X_c(t) := X(t) - X([1, \infty), t) - \sum_{n=0}^{\infty} \Big[X(\Delta_n, t) - \mathbf{E} X(\Delta_n, t) \Big].$$

The series on the right converges with probability 1, uniformly, as long as

$$\sum_{n=0}^{\infty} \int_{\Delta_n} |x|^2 \, \pi(dx, t) < \infty,$$

which implies that $X_c(t)$ has a.s. continuous sample paths (all discontinuities have been removed from $X(t)$). Thus, we arrive at a general random integral representation of an II process:

$$X(t) = X_c(t) + \int_{|x| \le 1} x \big(N(dx, t - \pi(dx, t)) + \int_{|x| > 1} x \, N(dx, t). \tag{5.7}$$

Since $X_c(t)$ is a II process with a.s. continuous paths, it is necessarily a generalized Brownian motion process (up to a constant) so that we arrive at the following representation of the characteristic function of a general II process:

$$\mathbf{E} \exp[iuX(t)] = \exp\Big[iua(t) - \frac{b(t)}{2} u^2 + \int_{|x| \le 1} \Big(e^{iux} - 1 - iux \Big) \pi(dx, t)$$

$$+ \int_{|x| > 1} \Big(e^{iux} - 1 \Big) \pi(dx, t) \Big]. \tag{5.8}$$

The above representation is known as the *Lévy-Khinchine formula*.

Example 5.4 (Back to α-stable processes). *Consider the special case of the Lévy-Khinchine formula with the measure*

$$\pi(dx, t) = t \frac{dx}{|x|^{\alpha+1}}, \qquad 0 < \alpha < 2,$$

and $a(t) = b(t) = 0$. Clearly,

$$\int_{0 < |x| \le 1} x^2 \pi(dx, t) < \infty, \quad \text{and} \quad \int_{|x| > 1} \pi(dx, t) < \infty,$$

so the corresponding process $X(t)$ has stationary and independent increments and the characteristic function

$$\mathbf{E} \exp[iuX(t)] = \exp t \Big[\int_{0 < |x| \le 1} \Big(e^{iux} - 1 - iux \Big) \frac{dx}{|x|^{\alpha+1}} + \int_{|x| > 1} \Big(e^{iux} - 1 \Big) \frac{dx}{|x|^{\alpha+1}} \Big].$$

$$= \exp\Big[2t \int_0^\infty (\cos(ux) - 1) \frac{dx}{|x|^{\alpha+1}} \Big] = e^{-Ct|u|^\alpha},$$

where the constant

$$C = 2 \int_0^\infty (1 - \cos y) \frac{dy}{y^{\alpha+1}}.$$

In addition to properties of α-stable processes discussed in Chapter 4, such as self-similarity,

$$X(ct) \overset{d}{=} c^{1/\alpha} X(t),$$

and stability,

$$aX + bY \overset{d}{=} (a^\alpha + b^\alpha)^{1/\alpha} X$$

for independent standard α-stable random variables X and Y, let us observe that since

$$\int_{0<|x|\leq 1} \frac{dx}{|x|^{\alpha+1}} = \infty$$

the α-stable process is not a pure jump process, that is its sample paths are not piece-wise constant a.s. A selection of sample path of such processes with different values of α is shown in Figure 5.2.

Remark 5.3 (Asymmetric α-stable processes). The general, possibly asymmetric α-stable process $X(t)$ with stationary increments has the Lévy measure

$$\pi(dx, t) = t \left(\frac{c^-}{|x|^{1+\alpha}} I_{(-\infty,0)}(x) \, dx + \frac{c^+}{x^{1+\alpha}} I_{(0,\infty)}(x) \, dx \right).$$

For various choices of the constants c^- and c^+, this leads to a family of processes with characteristic functions of the form, for $0 < \alpha < 2, \alpha \neq 1$,

$$\mathbf{E} \exp[iuX(t)] = \exp t \left[i\mu u - \sigma^\alpha |u|^\alpha \left(1 - i\beta \operatorname{sign}(u) \tan\left(\frac{\alpha\pi}{2} \right) \right) \right]$$

which contains four parameters: α – usually called the index of stability, $\beta \in [-1, 1]$ – the skewness parameter, $\sigma > 0$ – the scale parameter (playing the role similar to the standard deviation for Gaussian distribution), and the shift $\mu \in \mathbf{R}$.

For $\alpha = 1$, one obtains the usual (possibly shifted and skewed) Cauchy process., with the characteristic function

$$\mathbf{E} \exp[iuX(t)] = \exp t \left[i\mu u - \sigma |u| \left(1 + i\beta \frac{2}{\pi} \operatorname{sign}(u) \log u \right) \right].$$

Note that in the case of the totally asymmetric α-stable processes with $\alpha = 1/2$ and the Lévy measure

$$\pi(dx) = \frac{\sqrt{\pi}}{2x^{3/2}} \, dx, \qquad \text{for} \qquad x > 0,$$

and vanishing on the negative half-line, which generates the characteristic function

$$\mathbf{E} \exp[iuX(t)] = \exp t \left[-|u|^{1/2} e^{\pm i\pi/4} \right],$$

with the sign $+$ in the positive half-line $u > 0$, and $-$ on the negative half-line, the p.d.f can also be explicitly calculated, and

$$f_{X(1)}(x) = \frac{1}{\sqrt{2\pi} x^{3/2}} e^{-1/(2x)}, \qquad \text{for} \qquad x > 0,$$

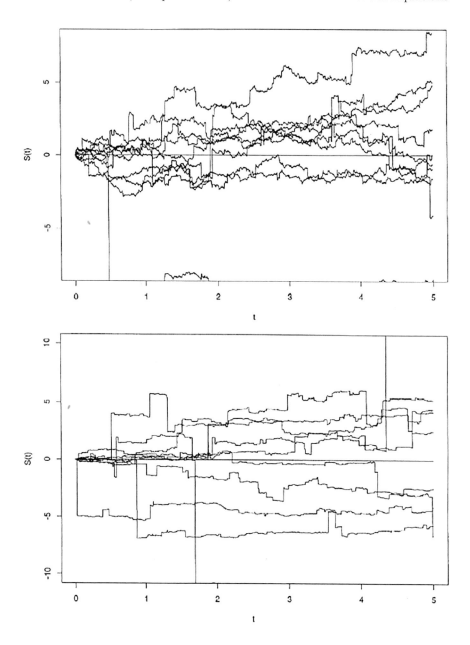

Figure 5.2 Sample paths of α-stable processes for $\alpha = 1.5$ (a), and $\alpha = 1$ (b). Compare them with sample paths in Figure 2.3. for Brownian motion, i.e. the two-stable process.

and 0 on the negative half-line. Observe that this distribution is exactly the distribution of the Brownian motion's hitting time of level 1 discussed in Section 2.5.

Remark 5.4 (Alternative formulations of the Lévy-Khinchine formula).
Sometimes it is more convenient to write the expression (5.8) in a different form.

Here are two alternatives:

$$\mathbf{E}\exp[iuX(t)] = \exp\left[iua(t) - \frac{b(t)}{2}u^2 + \int_{\mathbf{R}\backslash\{0\}}\left(e^{iux} - 1 - iuxI_{(0,1]}(x)\right)\pi(dx,t)\right]$$

and

$$\mathbf{E}\exp[iuX(t)] = \exp\left[iua(t) - \frac{b(t)}{2}u^2 + \int_{\mathbf{R}\backslash\{0\}}\left(e^{iux} - 1 - \frac{iux}{1+x^2}\right)\frac{1+x^2}{x^2}\rho(dx,t)\right]$$

In the first case, the Lévy measure π must satisfy the integrability condition $\int(1 \wedge x^2)\,\pi(dx) < \infty$, but in the second case only, the finiteness of measure ρ is required, so that, in the particular case of stationary increments one can write the Lévy-Khinchine formula in the form

$$\mathbf{E}\exp[iuX(t)] = \exp\left[iua - \frac{b}{2}u^2 + \int_{\mathbf{R}}\left(e^{iux} - 1 - \frac{iux}{1+x^2}\right)\frac{1+x^2}{x^2}\,dR(x)\right]$$

where $R(x) = \rho(-\infty, x]$ is an arbitrary increasing and bounded function equal to 0 at minus infinity.

Remark 5.5. Vector-valued s.i.i. processes and multidimensional Lévy-Khinchine formula. The definition of a s.i.i. vector -valued process $\vec{X}(t), t \geq 0$, with values in $\mathbf{R}^d$ is analogous to the one-dimensional case. In the general case, the characteristic function is of the form

$$\mathbf{E}\exp\left[i\langle\vec{u}, \vec{X}(t)\rangle\right] = \exp t\left[i\langle\vec{u}, \vec{a}\rangle - \frac{1}{2}\langle B\vec{u}, \vec{u}\rangle\right.$$
$$\left. + \int_{\mathbf{R}^d\backslash\{0\}}\left(e^{i\langle\vec{u},\vec{v}\rangle} - 1 - \langle\vec{u},\vec{v}\rangle I_{\|\vec{x}\|\leq 1}(x)\right)\pi(dx,t)\right]$$

where $\vec{a} \in \mathbf{R}^d$, the matrix B is non-negative definite, symmetric, and real-valued, and the Lévy measure π on R^d is σ-finite and satisfying the condition

$$\int_{\mathbf{R}^d\backslash\{0\}}(1 \wedge \|\vec{x}\|^2)\,\pi(dx) < \infty.$$

Remark 5.6. Infinite divisibility of distributions of Lévy processes. Finally, let us observe that the distributions of s.i.i. processes are infinitely divisible. Indeed, for any $t > 0, n = 2, 3, \ldots,$, we can write

$$B(t) = \left[B\left(\frac{t}{n}\right) - B(0)\right] + \cdots + \left[B\left(\frac{nt}{n}\right) - B\left(\frac{(n-1)t}{n}\right)\right],$$

where the n summands on the right are independent and identically distributed.

CHAPTER 6

Stochastic Integrals for Brownian Motion and General Lévy Processes

6.1 WIENER RANDOM INTEGRAL

Random measure (orthogonally scattered) on $([0,1], \mathcal{B})$: A mapping

$$\mathcal{B} \ni A \longmapsto M(A) \in L^2(\Omega.\mathcal{F}, \mathbf{P}) \tag{6.1}$$

such that

$$\mathbf{E}M(A) = 0$$
$$M(A \cup B) = M(A) + M(B), \qquad \text{if} \quad A \cap B = \emptyset,$$

and

$$\mathbf{E}M(A) \cdot M(B) = 0, \qquad \text{if} \quad A \cap B = \emptyset.$$

Brownian motion generates a random measure by the extension of the formula

$$M((a,b]) = B_b - B_a.$$

So does the symmetrization of the Poisson process.

Construction of the Wiener random integral. Start with simple functions

$$f(t) = \sum_i a_i I(A_i),$$

where A_i's are disjoint and define

$$\int_0^1 f(t) M(dt) = \sum_i a_i M(A_i). \tag{6.2}$$

DOI: 10.1201/9781003216759-6

Then $\mathbf{E} \int_0^1 f(t) M(dt) = 0$, and we have the isometry property,

$$\mathbf{E} \left| \int_0^1 f(t) M(dt) \right|^2 = \int_0^1 |f(t)|^2 m(dt) \qquad (6.3)$$

where $m(A) = E|M(A)|^2$ is the control measure. The isometry permits extension of the random integral to all functions $f \in L^2([0,1], \mathcal{B}, m)$.

$$X_n = \int_0^1 e^{int} M(dt),$$

gives a representation of second-order weakly stationary processes.

6.2 ITÔ'S STOCHASTIC INTEGRAL FOR BROWNIAN MOTION

Trying to extend the definition of the Wiener integral to the case of stochastic integrands,

$$\int_0^t f(s, \omega) dB_s(\omega), \qquad (6.4)$$

poses new difficulties. The motivation is to make rigorous the differential equations with the additive white noise

$$\frac{dX_t}{dt} = b(t, X_t) + \sigma(t, X_t) \cdot \dot{B}_t,$$

which, on the face value, does not make any sense because of nondifferentiability of the Brownian motion trajectories. Therefore, the equation is reinterpreted as an integral equation

$$X_t = X_0 + \int_0^t b(s, X_s) \, ds + \int_0^t \sigma(s, X_s) \, dB_s,$$

thus necessitating the development of the rigorous definition and theory for the stochastic integral (6.4).

Trying to pursue the Riemann-style construction produces difficulties. Again start with simple integrands, this time taking random values on disjoint intervals. However, to get the isometry property, we will require a certain structure of the statistical dependence between f_t and B_t.

Example 6.1 (To anticipate or not to anticipate - that is the question).
Take $f(t) = B(t)$, and consider the integral $\int B_t \, dB_t$ as a limit of the integrals of elementary (step) functions $e(s, \omega)$ corresponding to a partition

$$0 = t_0 < t_1 < \cdots < t_n = t$$

(a) Itô's Choice. If the step functions are selected to be

$$e(s, \omega) = \sum_{j=0}^{n-1} B_{t_j} \cdot I_{[t_j, t_{j+1})}(s)$$

then

$$\mathbf{E}\left(\int_0^t e(s,\omega)dB_t(\omega)\right) = \sum_{j=0}^{n-1}\mathbf{E}B_{t_j}(B_{t_{j+1}} - B_{t_j}) = \sum_{j=0}^{n-1}\mathbf{E}B_{t_j}\mathbf{E}(B_{t_{j+1}} - B_{t_j}) = 0,$$

and, by taking conditional expectations,

$$\mathbf{E}\left(\int_0^t e(s,\omega)dB_s(\omega)\right)^2 = \sum_{j=0}^{n-1}\sum_{i=0}^{n-1}\mathbf{E}B_{t_j}B_{t_i}(B_{t_{j+1}} - B_{t_j})(B_{t_{i+1}} - B_{t_i})$$

$$= \sum_{i=0}^{n-1}\mathbf{E}B_{t_i}^2\mathbf{E}(B_{t_{i+1}} - B_{t_i})^2 = \approx t^2/2$$

because for the off-diagonal terms, say $j < i$,

$$\mathbf{E}B_{t_j}B_{t_i}(B_{t_{j+1}}-B_{t_j})(B_{t_{i+1}}-B_{t_i}) = \mathbf{E}B_{t_j}B_{t_i}(B_{t_{j+1}}-B_{t_j})\mathbf{E}\big((B_{t_{i+1}}-B_{t_i})\,\big|\,\mathcal{F}_{t_i}\big) = 0,$$

where $\mathcal{F}_t$ is the sigma-field generated by $B_s, s \leq t$.

(b) Bad Choice. However, if the step functions are selected to be

$$e(s,\omega) = \sum_{j=0}^{n-1}B_{t_{j+1}}\cdot I_{[t_j,t_{j+1})}(s)$$

then

$$\mathbf{E}\left(\int_0^t e(s,\omega)dB_s(\omega)\right) = \sum_{j=0}^{n-1}\mathbf{E}B_{t_{j+1}}(B_{t_{j+1}} - B_{t_j})$$

$$= \sum_{j=0}^{n-1}\mathbf{E}(B_{t_{j+1}} - B_{t_j})^2 = t,$$

Therefore, the result is very different from (a). The choice (a) leads to the definition of Itô's stochastic integral.

(b) Stratonovich's Choice. However, if the step functions are selected to be

$$e(s,\omega) = \sum_{j=0}^{n-1}B_{\frac{t_{j+1}+t_j}{2}}\cdot I_{[t_j,t_{j+1})}(s)$$

then

$$\mathbf{E}\left(\int_0^t e(s,\omega)dB_t(\omega)\right) = \sum_{j=0}^{n-1}\mathbf{E}B_{\frac{t_{j+1}+t_j}{2}}(B_{t_{j+1}} - B_{t_j})$$

$$= \sum_{j=0}^{n-1}\mathbf{E}B_{\frac{t_{j+1}+t_j}{2}}\left((B_{t_{j+1}} - B_{\frac{t_{j+1}+t_j}{2}}) + (B_{\frac{t_{j+1}+t_j}{2}} - B_{t_j})\right)$$

$$= \sum_{j=0}^{n-1}\mathbf{E}(B_{\frac{t_{j+1}+t_j}{2}} - B_{t_j})^2 \approx \frac{t}{2},$$

Therefore, the result is very different from (a) and (b). The choice (a) leads to the definition of Stratonovich's stochastic integral.

Definition 6.1. An integrand process $f(t, \omega), t \in [0, 1]$, is said to be adapted (nonanticipating) if

$$f_t \text{ meas } \mathcal{F}_t := \sigma(B_s, s \leq t) \quad \forall t \in [0, 1].$$

An integrand process $e(t, \omega), t \in [0, 1]$ is said to be elementary if it is of the form

$$e(t, \omega) = \sum_j e_j(\omega) I_{[t_j, t_{j+1})}(t)$$

with e_j meas $\mathcal{F}_{t_j} \; \forall j$.

Definition 6.2. Itô's (nonanticipating) integral of an elementary process:

$$\int_0^t e_t dB_t = \sum_j e_j(B_{t_{j+1}} - B_{t_j}) \equiv \sum_j e_j \Delta B_{j+1}.$$

Lemma 6.1. *We have*

$$\mathbf{E} \int e_t \, dB_t = 0,$$

and

$$\mathbf{E} \left(\int e_t \, dB_t \right)^2 = \mathbf{E} \int e_t^2 \, dt.$$

Proof. The expectation

$$\mathbf{E} \int e_t \, dB_t = \mathbf{E} \sum_j e_j \Delta B_{j+1} = \mathbf{E} \sum_j \mathbf{E}[e_j \Delta B_{j+1} | \mathcal{F}_j]$$

$$= \mathbf{E} \sum_j e_j \mathbf{E}[\Delta B_{j+1} | \mathcal{F}_j] = 0,$$

and the second moment

$$\mathbf{E} \left(\int e_t \, dB_t \right)^2 = \mathbf{E} \left(\sum_j e_j \Delta B_{j+1} \cdot \sum_i e_i \Delta B_{i+1} \right)$$

$$= \sum_j \mathbf{E}(e_j)^2 E(\Delta B_{j+1})^2 = \mathbf{E} \int e_t^2 \, dt.$$

□

Now we can extend the definition to a larger class of nonanticipating integrand processes:

Theorem 6.1. *For each nonanticipating process f_t such that*

$$\mathbf{E} \int_0^t |f_s|^2 \, ds < \infty \tag{6.5}$$

there exists a sequence of $e_n(t)$ of elementary processes such that, as $n, m \to \infty$,

$$\mathbf{E}\int_0^t |f - e_n|^2 \, ds \to 0, \quad and \quad \mathbf{E}\left(\int_0^t (e_m - e_n) \, dB_s\right)^2 \to 0. \qquad (6.6)$$

Then

$$\int_0^t f_s \, dB_s := (L_2) - \lim_{n \to \infty} \int_0^t e_n(s) \, dB_s. \qquad (6.7)$$

For nonanticipating integrands, we also have the L_2- isometry,

$$\mathbf{E}\left|\int f_t \, dB_t\right|^2 = \mathbf{E}\int |f_t|^2 \, dt. \qquad (6.8)$$

The Lebesgue measure is the control measure for the Brownian motion.

Proof. In view of the Lemma, it suffices to show the existence of the desired approximating sequence of elementary processes. This can be done in three steps.

Step 1. Define $h_n(t, \omega) = \min(f(t, \omega), n)$. In view of the dominated convergence theorem,

$$\mathbf{E}\int_0^t (f - h_n)^2 \, ds \to 0, \quad \text{as} \quad n \to \infty.$$

Step 2. For a bounded $h(t, \omega)$, take

$$g_n(t, \omega) = \int_0^t \phi_n(s - t) h(s, \omega) \, ds,$$

where ϕ_n is the approximate identity sequence of functions located to the left of 0 (Figure 6.1).

Processes g_n are nonanticipating and have bounded continuous trajectories, and

$$\mathbf{E}\int_0^t \left(g_n(s) - h(s)\right)^2 \, ds \to 0, \quad \text{as} \quad n \to \infty.$$

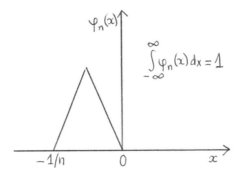

Figure 6.1 Approximate identity kernel preserving the nonanticipating structure.

Step 3. Since $g_n(t, \omega)$ have continuous trajectories, they can be approximated by elementary processes. □

Itô's integrals also have the martingale structure:

$$\mathbf{E}\left(\int_0^s f_t dB_t \,\Big|\, \mathcal{F}_u\right) = \int_0^u f_t \, dB_t, \qquad u < s, \tag{6.9}$$

This property assures the following maximal inequality:

$$\mathbf{P}\left(\sup_{0 \le t \le u} \left|\int_0^t f_s \, dB_s\right| > x\right) \le \frac{\mathbf{E}\int_0^u |f_s|^2 \, ds}{x^2}, \qquad \forall u \in [0, 1], x > 0, \tag{6.10}$$

6.3   AN INSTRUCTIVE EXAMPLE

As an instructive example, let us try to find the stochastic integral $\int_0^t B_s \, dB_s$. The answer is not $B_t^2/2$ because

$$\mathbf{E}\int_0^t B_s \, dB_s = 0, \qquad \text{while} \qquad \mathbf{E}B_t^2/2 = t/2.$$

It turns out that

$$\int_0^t B_s dB_s = \frac{1}{2}(B_t^2 - t)$$

To prove this directly from the definition, let us take the approximating elementary processes corresponding to the partition $0 = t_0 < \cdots < t_n = t$,

$$e_s = \sum_{j=1}^n B_j I_{[t_j, t_{j+1})}, \qquad B_j = B_{t_j},$$

which, with mesh $\max_j |t_{j+1} - t_j| \to 0$, approximates B_t in L^2:

$$\mathbf{E}\left(\int_0^t (e_s - B_s)^2 \, ds\right) = \mathbf{E}\left(\sum_{j=1}^n \int_{t_j}^{t_{j+1}} (B_j - B_s)^2 \, ds\right)$$

$$= \left(\sum_{j=1}^n \int_{t_j}^{t_{j+1}} (s - t_j) \, ds\right) = \frac{1}{2}\sum_{j=1}^n (t_{j+1} - t_j)^2 \, ds \to 0$$

Now, expanding,

$$\mathbf{E}\left(\frac{1}{2}(B_t^2 - t) - \sum_j B_j \Delta B_{j+1}\right)^2$$

$$= \frac{3}{4}t^2 - \frac{1}{2}t^2 + \frac{1}{4}t^2 - \mathbf{E}B_t^2 \sum_j B_j \Delta B_{j+1} + t\mathbf{E}\sum_j B_j \Delta B_{j+1} + \mathbf{E}\left(\sum_j B_j \Delta B_{j+1}\right)^2$$

$$= \frac{1}{2}t^2 - \sum_j t_j \Delta t_{j+1} \to 0,$$

as $\max_j |t_{j+1} - t_j| \to 0$, because the fourth term is

$$-\sum_j \mathbf{E} B_j \Delta B_{j+1} \Big((B_t - B_{j+1}) + \Delta B_{j+1} + B_j \Big)^2 = -2 \sum_j t_j \Delta t_{j+1},$$

the fifth term is 0, and the sixth term is $\sum_j t_j \Delta t_{j+1}$.

Is there a more general formula behind the above calculations? We shall answer this question in the next section.

6.4 ITÔ'S FORMULA

Because the integrator in the Itô's Brownian integral has infinite variation, the standard rules of calculus do not apply. In particular, the change-of-variables formula (chain rule), called the Itô's formula, takes a different form:

Theorem 6.2. *Let*

$$X_t = X_0 + \int_0^t u(s,\omega)ds + \int_0^t v(s,\omega)dB_s \tag{6.11}$$

be a stochastic integral process with nonanticipating processes u, and v, such that

$$\mathbf{P}\left(\int_0^t v^2(s)ds < \infty, \forall t > 0\right) = 1,$$

and

$$\mathbf{P}\left(\int_0^t |u(s)|ds < \infty, \forall t > 0\right) = 1.$$

If $g(t,x)t \geq 0, x \in \mathbf{R}$ is twice differentiable then the process

$$Y_t = g(t, X_t)$$

is also a stochastic integral and

$$Y_t = Y_0 + \int_0^t \frac{\partial g}{\partial s}(s, X_s)ds + \int_0^t \frac{\partial g}{\partial x}(s, X_s)dX_s + \frac{1}{2}\int_0^t \frac{\partial^2 g}{\partial x^2}(s, X_s)(dX_s)^2, \tag{6.12}$$

where $(dX_t)^2$ is calculated according to the following rules:

$$dtdt = 0, \quad dtdB_t = 0, \quad dB_t dB_t = dt.$$

Proof. Itô's formula is equivalent to the following statement:

$$g(t, X_t) = g(0, X_0) + \int_0^t \left(\frac{\partial g}{\partial s} + u\frac{\partial g}{\partial x} + \frac{1}{2}v^2\frac{\partial^2 g}{\partial x^2}\right)ds + \int_0^t v\frac{\partial g}{\partial x}dB_s \tag{6.13}$$

It suffices to establish (7.3) four bounded elementary processes, u, v, and bounded g with bounded first derivative in t and second derivative in x. In this case, by Taylor's theorem,

$$g(t, X_t = g(0, X_0) + \sum_j \Delta g(t_j, X_j)$$

$$= g(0, X_0) + \sum_j \frac{\partial g}{\partial t} \Delta t_j + \sum_j \frac{\partial g}{\partial x} \Delta X_j$$

$$+ \frac{1}{2} \sum_j \frac{\partial^2 g}{\partial t^2} (\Delta t_j)^2 + \sum_j \frac{\partial^2 g}{\partial t \partial x} \Delta t_j \Delta X_j + \frac{1}{2} \sum_j \frac{\partial^2 g}{\partial x^2} (\Delta X_j)^2 + \sum_j R_j,$$

where

$$\Delta g(t_j, X_j) = g(t_{j+1}, X_{j+1}) - g(t_j, X_j), \qquad \text{and} \qquad R_j = o(|\Delta t_j|^2 + |\Delta X_j|^2).$$

If $\max_j \Delta t_j \to 0$, then

$$\sum_j \frac{\partial g}{\partial t} \Delta t_j \longrightarrow \int_0^t \frac{\partial g}{\partial s}(s, X_s)\, ds,$$

$$\sum_j \frac{\partial g}{\partial x} \Delta X_j \longrightarrow \int_0^t \frac{\partial g}{\partial x}(s, X_s)\, dX_s,$$

and

$$\sum_j \frac{\partial^2 g}{\partial x^2} (\Delta X_j)^2 = \sum_j \frac{\partial^2 g}{\partial x^2} u_j^2 (\Delta t_j)^2 + 2 \sum_j \frac{\partial^2 g}{\partial x^2} u_j v_j \Delta t_j \Delta B_j + \sum_j \frac{\partial^2 g}{\partial x^2} v_j^2 (\Delta B_j)^2.$$

The first two terms on the right-hand side converge to 0, and

$$\mathbf{E} \left(\sum_j \frac{\partial^2 g}{\partial x^2} v_j^2 (\Delta B_j)^2 - \int_0^t \frac{\partial^2 g}{\partial x^2} v^2\, ds \right)^2 \longrightarrow 0.$$

$\square$

Example 6.2.

(a) *In view of Itô's formula, taking* $g(t, x) = x^2$, *we have*

$$d(B_t)^2 = 2 B_t dB_t + dt,$$

so

$$\int_0^t B_s\, dB_s = \frac{1}{2}(B_t^2 - t).$$

(b) *Taking* $g(t, x) = tx$, *we have*

$$d(t B_t) = t\, dB_t + dt\, B_t,$$

so

$$\int_0^t s\, dB_s = t B_t - \int_0^t B_s\, ds.$$

6.5   MARTINGALE PROPERTY OF ITÔ INTEGRALS

Let us begin by recalling the definition of a martingale.

Definition 6.3. Let $\mathcal{F}_t, \in [0, T]$, be a family of ascending sub-σ-fields (i.e., a filtration) of the σ-field of random events. A stochastic process M_t is called a *martingale* with respect to filtration $\mathcal{F}_t, t \in [0, T]$, if it satisfies the following two conditions:

(i) For each $t \in [0, T]$, M_t is measurable with respect to $\mathcal{F}_t$, and $\mathbf{E}|M_t| < \infty$, and

(ii) For each pair $s \le t$ in $[0, T]$, the conditional expectation $\mathbf{E}(M_t|\mathcal{F}_s) = M_s$.

Example 6.3. *The Brownian motion process $B_t, t \ge 0$, is a martingale with respect to the natural filtration $\mathcal{F}_t = \sigma\{B_s, s \le t\}$. Indeed, the measurability condition is satisfied because of the definition of the filtration, and, in view of the Cauchy-Schwartz inequality,*

$$\mathbf{E}|B_t| \le (\mathbf{E}B_t^2)^{1/2} = \sqrt{t} < \infty.$$

Also, condition (ii) is satisfied since

$$\mathbf{E}(B_t|\mathcal{F}_s) = \mathbf{E}(B_t - B_s|\mathcal{F}_s) + \mathbf{E}(B_s|\mathcal{F}_s) = 0 + B_s.$$

Theorem 6.3. *If process $f(t, \omega), t \in [0, T]$ is nonanticipating with*

$$\mathbf{E} \int_T |f(t)^2 dt < \infty,$$

then

$$I_t = \int_0^t f(s)\, dB_s, \qquad t \in [0, T],$$

is a martingale with respect to the natural filtration $\mathcal{F}_t = \sigma\{B_s, s \le t\}$.

Proof. Choose elementary processes $E^{(n)} \longrightarrow f$ in L^2, as $n \to \infty$, and define

$$I_t^{(n)} = \int_0^t e^{(n)}(s)\, dB_s, \qquad t \in [0, T],$$

Clearly, $I_t^{(n)}$ is a martingale with continuous sample paths, because for $s \le t$

$$\mathbf{E}(I_t^{(n)}|\mathcal{F}_s) = \mathbf{E}\left[\int_0^s e^{(n)}\, dB + \int_s^t e^{(n)}\, dB \,\bigg|\, \mathcal{F}_s \right] = \int_0^s e^{(n)}\, dB = I_s^{(n)}.$$

and the martingale property follows by continuity. As a result, $I_t^{(n)} - I_t^{(m)}$ is also an $\mathcal{F}_t$-martingale and, in view of *Doob's martingale inequality*

$$\mathbf{P}\left(\sup_{0 \le t \le T} |M_t| \ge \lambda \right) \le \frac{\mathbf{E}|M_T|^p}{\lambda^p}, \qquad \lambda \ge 0, p \ge 1,$$

and the L^2 isometry property of the stochastic integral, for $p = 2$, we get

$$\mathbf{P}\left(\sup_{0 \leq t \leq T} ||I_t^{(n)} - I_t^{(m)}| > \epsilon\right) \leq \frac{2}{\epsilon^2}\mathbf{E}\left[\int_0^T (e^{(n)} - e^{(m)})^2\, ds\right] \to 0, .$$

as $n, m \to \infty$.

Therefore, there exists a sequence $n_k \uparrow \infty$, such that

$$\mathbf{P}\left(\sup_{0 \leq t \leq T} ||I_t^{(n_k+1)} - I_t^{(n_k)}| > 2^{-k}\right) < 2^{-k},$$

for all $k = 1, 2, \ldots$, and by the Borell–Cantelli Lemma, with probability 0, for infinitely many k's

$$\sup_{0 \leq t \leq T} |I_t^{(n_k+1)} - I_t^{(n_k)}| > 2^{-k},$$

and, consequently, there exists a random variable $K(\omega)$, such that, for all $k > K(\omega)$,

$$\sup_{0 \leq t \leq T} |I_t^{(n_k+1)} - I_t^{(n_k)}| < 2^{-k},$$

almost surely. As a result, $I_t^{(n_k)}$ converges uniformly on $[0, T]$, a.s., which implies the continuity of $\lim_{k \to infty} I_t^{(n_k)} = I_t$ (in $L^{(dP)}$), for all t. Thus, it is always possible to choose a continuous modification of I_t. □

Corollary 6.1. *As a consequence of the martingale property for stochastic integrals, and the L^2 isometry property, we obtain the following version of Doob's inequality*

$$\mathbf{P}\left(\sup_{0 \leq t \leq T} |\int_0^T f_s\, dB_s| \geq \lambda\right) \leq \frac{1}{\lambda^2}\mathbf{E}\int_0^T f_s^2\, ds.$$

6.6 WIENER AND ITÔ-TYPE STOCHASTIC INTEGRALS FOR α-STABLE MOTION AND GENERAL LÉVY PROCESSES

For the α-stable motion, $X_t, \alpha < 2$, with the second moment infinite, we cannot define the Wiener integral

$$\int_0^t f_s dX_s(\omega) \tag{6.14}$$

through the method introduced above for the Brownian motion. The way around this difficulty is to rely on the knowledge of the characteristic function of X_t. Then for the approximation of f_t by a step function e_t, we get

$$\mathbf{E}\exp\left(\sum_j e_j \Delta X_{j+1}\right) = \exp\left(\sum |e_j|^\alpha \Delta t_{j+1}\right) \to \exp\left(\int_0^t |f_s|^\alpha\, ds\right),$$

as the partition's mesh tends to zero. Thus, the integral (6.11) is well defined for any function $f \in L^\alpha[0, 1]$.

Existence of the stochastic Itô-type integral

$$\int_0^t f_s(\omega) dX_s(\omega) \tag{6.15}$$

can be demonstrated in the case of the integrand process $f_t(\omega)$ having trajectories in $L^\alpha[0,1]$ a.s. (see, e.g., Kwapien and Woyczynski (1992), for general Lévy integrators, and Rosinski and Woyczynski (1986) for the α-stable case.)

CHAPTER 7

Itô Stochastic Differential Equations

7.1 DIFFERENTIAL EQUATIONS WITH NOISE

In this section, we provide a few examples of physical, economical, or engineering situations where considering dynamical systems described by differential equation in the presence of "random noise" is necessary and justifies at the intuitive level the formal development of stochastic differential equation, which will be presented in the following sections.

Example 7.1 (Population growth model). *Consider a population growing in a random environment affected by the weather, climate, and other environmental conditions. The rate of growth, $r(t)$, is, in general, time-dependent, but also dependent on the environmental, also time-dependent "noise". So the size of the population, $N(t)$, can be described by the differential equation*

$$\frac{dN(t)}{dT} = \Big(r(t) + \text{``noise''}(t) \Big) N(t)$$

with the initial condition $N(0) = N_0$ describing the original size of the population.

Example 7.2 (Electrical circuit with thermal noise). *The current intensity $itor(t)$ in an electrical circuit (shown in Figure 7.1) involving a capacitor with capacitance C, a resistor with resistivity R, and inductor with inductance L, is described by the integro-differential equation*

$$L\frac{di(t)}{dt} + Ri(t) + \frac{1}{C} \int_0^t i(s)\, ds = V(t),$$

where $U(t)$ is the voltage applied to the circuit. In many situations, the voltage $U(t)$ has a random "noise" component, which can be the results of transmission through a random medium or the thermal noise of the electrons moving around. In this case, after differentiation, the above integro-differential equation can be rewritten in the

DOI: 10.1201/9781003216759-7

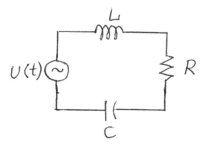

Figure 7.1 An example of electronic circuit involving a capacitor, resistor and inductor, in the presence of external voltage.

form of a second-order differential equation

$$Li''(t) + Ri'(t) + \frac{1}{C}i(t) = V'(t) + \text{``}noise\text{''}(t).$$

Example 7.3 (Filtering and prediction). *Assume we observe a stochastic $Q(t)$ generated by a random mechanism at times $t_1 < t_2 < \ldots t_n$. The observations are transmitted through channel in the presence of "noise" , so the result of observations is a sequence of random variables*

$$Z(t_k) = Q(t_k) + \text{``}noise\text{''}(t_k), \qquad k = 1, 2, \ldots, n,$$

There are two questions related to this situation:

(i) *How do we recover $Q(t_k), k = 1, 2, \ldots, n$, from observations of $Z(t_k), k = 1, 2, \ldots, n$ (the filtering problem)?*

(ii) *How do we estimate $Q(t), t > t_n$, based on our observations (the prediction problem)?*

These problems and be stated in the language of stochastic differential equations and the answer is provided by the so-called Kalman-Bucy filter construction to be discussed later.

Example 7.4 (Stochastic control). *The macroeconomic question "How much should a nation save?" can be formulated as a problem of optimization for a system of differential equation with "noise". Denote by $L(t)$ the size of the labor force, $K(t)$–the capital, $P(t)$ – production rate, and $C(t)$ — consumption rate at time t. With the labor force dynamics described by the population growth equation from Example 7.1, the set of generally accepted equations tying together the above quantities are as follows:*

$$\frac{dL(t)}{dt} = (r(t) + \text{``}noise\text{''}(t))L(t),$$

$$\frac{dK(t)}{dt} = P(t) - C(t),$$

and

$$P(t) = AK^{\alpha}(t)L^{\beta}(t),$$

for some values of the constants A, α, and β. Given the "utility function" $U(c)$, with the "utiliy" $U(c)\Delta t$ interpreted as consumption of goods at rate c in time interval Δt, and the "bequest" function ψ, the optimization problem can be phrased as follows: Determine the consumption rate $C(t)$ which maximizes the "total utility"

$$\mathbf{E}\left[\int_0^T U(C(t))e^{-\rho t}dt\right] + \psi(K(T)).$$

7.2 STOCHASTIC DIFFERENTIAL EQUATIONS: BASIC THEORY

Examples of differential equations with "noise" given in the previous section can be written in the general form

$$\frac{dX_t}{dt} = b(t, X_t) + \sigma(t, X_t)\dot{B}_t,$$

where, in the special case, $\dot{B}_t$ is the so-called white noise, .i.e., the derivative of the Brownian motion discussed in earlier chapters. However, since Brownian motion is not differentiable, the rigorous approach calls for the interpretation of the above stochastic equation as Itô's stochastic integral equation

$$X_s = X_0 + \int_0^s b(t, X_t)dt + \int_0^s \sigma(t, X_t)dB_t, \tag{7.1}$$

which is traditional written in the form of the stochastic differential equation

$$dX_t = b(t, X_t)dt + \sigma(t, X_t)dB_t. \tag{7.2}$$

Note that using Ito's formula, we can study transformations of the above SDEs. Indeed, consider a function $g(t, x)$ once continuously differentiable in t and twice continuous differentiable in x and monotone in x. There conditions guarantee, for each t, the existence of an inverse function $\hat{g}(t, x)$ such that

$$g(t, \hat{g}(t, x)) = x, \quad \text{and} \quad \hat{g}(t, g(t, x)) = x.$$

Now consider the new process

$$Y(t) = g(t, X(t)), \quad \text{so that} \quad X(t) = \hat{g}(t, Y(t)).$$

In view of Ito's formula, if $X(t)$ satisfies Equation (7.2) then

$$dY(t) = \left(\frac{\partial}{\partial t}g(t, X(t)) + b(t, X(t))\frac{\partial}{\partial x}g(t, X(t)) + \frac{1}{2}\sigma^2(t, X(t))\frac{\partial^2}{\partial^2 x}g(t, X(t))\right)dt$$

$$+ \sigma(t, X(t))\frac{\partial}{\partial x}g(t, X(t))\,dB(t).$$

Rephrasing the above equation, we can say that the process $Y(t)$ satisfies the SDE

$$dX_t = \hat{b}(t, X_t)dt + \hat{\sigma}(t, X_t)dB_t. \tag{7.3}$$

with

$$\hat{b}(t, x) = \frac{\partial}{\partial t}g(t, x) + b(t, x)\frac{\partial}{\partial x}g(t, x) + \frac{1}{2}\sigma^2(t, x)\frac{\partial^2}{\partial^2 x}g(t, x) \tag{7.4}$$

and

$$\hat{\sigma}(t, x) = \sigma(t, x)\frac{\partial}{\partial x}g(t, x). \tag{7.5}$$

Example 7.5. *Should we aim at making the coefficient in front of $dB(t)$ in (7.3) constant, say 1, that is demanding that*

$$\hat{g}(t, x) = \sigma(t, x)\frac{\partial}{\partial x}g(t, x) = 1,$$

we obtain

$$g(t, x) = \int_0^x \frac{dz}{\sigma(t, z)}.$$

which is valid for positive and continuously differentiable (in x) $\sigma(t, x)$.

Example 7.6. *Should we demand that the coefficient in (7.3) in front of dt disappear, the transforming function $g(t, x)$ has to satisfy the equation*

$$\frac{\partial}{\partial t}g(t, x) + b(t, x)\frac{\partial}{\partial x}g(t, x) + \frac{1}{2}\sigma^2(t, x)\frac{\partial^2}{\partial^2 x}g(t, x) = 0$$

In the special case when the coefficients b and σ are independent of t, the above equation becomes an ordinary differential equation,

$$b(x)g'(x) + \frac{1}{2}\sigma^2(x)g''(x) = 0,$$

with an explicit solution

$$g(x) = C_1 + C_2 \int_0^x \exp\left(-\int_0^y \frac{2b(z)}{\sigma^2(z)}dz\right)dy.$$

The first question in this context is: under what assumptions we can guarantee the existence and uniqueness of solutions for Equation (7.2). Below, we prove the basic theorem which provides an answer to the above question. We formulate it in the one-dimensional case, although the analogous multidimensional case can also be demonstrated[1]

---

[1] See, e.g., B. Oksendal, Stochastic Differential Equations, Sixth Edition, Springer, 2007.

Theorem 7.1. *If measurable real-valued coefficients $b(t,x)$, and $\sigma(t,x)$, of the equation*

$$dX_t = b(t, X_t)dt + \sigma(t, X_t)dB_t. \tag{7.6}$$

satisfy the growth condition,

$$|b(t,x)|^2 + |\sigma(t,x)|^2 \leq C^2(1 + |x|^2), \tag{7.7}$$

and the Lipschitz condition,

$$|b(t,x) - b(t,y)| + |\sigma(t,x) - b(t,y)| \leq C|x - y|, \tag{7.8}$$

for some constant C and D, and all $x, y \in \mathbf{R}$ and $t \in [0, T]$, and if the initial condition, $X(0) = X_0$, has finite variance and is independent of the σ-algebra generated by the Brownian motion $B_t, t \geq 0$, then the stochastic differential equation (7.2) has a solution which,

(i) *is continuous in the time variable t with probability 1 with $X(t) = X(0)$, for $t = 0$,*

(ii) *has bounded second moments, that is*

$$\sup_{0<t<T} \mathbf{E}|X_t|^2 < \infty \tag{7.9}$$

The solution is unique, that is, if $X_1(t)$, and $X_2(t)$, are two solutions of equation (7.6) then

$$\mathbf{P}\Big(\sup_{0<t<T} |X_1(t) - X_2(t)| = 0 \Big) = 1.$$

We precede the proof of the theorem by a what is often called

Gronwall Lemma. *If $\phi(t)$ and $\psi(t)$ are bounded measurable function satisfying the condition*

$$\phi(t) \leq \psi(t) + L \int_0^t \phi(s)\, ds,$$

then

$$\phi(t) \leq \psi(t) + L \int_0^t e^{L(t-s)} \psi(s)\, ds.$$

Proof. Observe that

$$\psi(t) + L \int_0^t e^{L(t-s)} \psi(s)\, ds - \phi(t) \equiv \gamma(t) \geq L \int_0^t \gamma(s)\, ds$$

$$L^2 \int_0^t \int_0^s \gamma(u)\, du\, ds = L^2 \int_0^t (t-u)\gamma(u)\, du \geq L^3 \int_0^t (t-u)\int_0^u \gamma(u)\, du\, ds$$

$$= L^3 \int_0^t \frac{1}{2}(t-s)^2 \gamma(s)\, ds \geq \cdots \geq \frac{1}{(n-1)!} L^n \int_0^t (t-s)^{n-1}\gamma(s)\, ds.$$

As $n \to \infty$, the last expression converges to 0, which gives the desired Gronwall inequality. $\qquad\square$

Now we return to the proof of uniqueness and existence of solutions of Equation (7.6) under assumptions (7.7–7.8). We'll also prove their properties (i) and (ii).

Proof. (**Proof of Theorem 7.1**). *Uniqueness:* Suppose we have two solutions $X_1(t)$, and $X_2(T)$ satisfying equation

$$X_i(t) = X(0) + \int_0^t b(s, X_i(s))ds + \int_0^t \sigma(s, X_i(s))dB(s), \quad i = 1, 2. \tag{7.10}$$

Then,

$$\mathbf{E}[X_1(t) - X_2(t)]^2 \tag{7.11}$$

$$= \mathbf{E}\left[\int_0^t [b(s, X_1(s)) - b(s, X_2(s))]ds + \int_0^t [\sigma(s, X_1(s)) - \sigma(s, X_2(s))]dB(s)\right]^2$$

$$\leq 2\mathbf{E}\left[\int_0^t [b(s, X_1(s)) - b(s, X_2(s))]ds\right]^2$$

$$+ 2\mathbf{E}\left[\int_0^t [\sigma(s, X_1(s)) - \sigma(s, X_2(s))]dB(s)\right]^2$$

$$\leq 2t\mathbf{E}\int_0^t [b(s, X_1(s)) - b(s, X_2(s))]^2 ds + 2\mathbf{E}\int_0^t [\sigma(s, X_1(s)) - \sigma(s, X_2(s))]^2 ds$$

$$\leq 2tK^2 \int_0^t \mathbf{E}[b(s, X_1(s)) - b(s, X_2(s))]^2 ds$$

$$+ 2K^2 \int_0^t \mathbf{E}[\sigma(s, X_1(s)) - \sigma(s, X_2(s))]^2 ds$$

$$\leq L \int_0^t \mathbf{E}|X_1(s) - X_2(s)|^2 ds,$$

where $L = 2(T + 1)K^2$. Now, in view of Gronwall's Lemma,

$$\mathbf{E}|X_1(t) - X_2(t)| = 0,$$

so that, for each $t \in [0, T]$, we have $\mathbf{P}[X_1(t) = X_2(t)] = 1$, which implies that for any countable subset $N \subset [0, T]$,

$$\mathbf{P}[\sup_{t \in N} |X_1(t) - X_2(t) = 0] = 1.$$

Finally, choosing N to be dense in the interval $[0, T]$, and utilizing the continuity of trajectories of $X_1(t)$, and $X_2(t)$, we get that

$$\mathbf{P}[\sup_{t \in [0,T]} |X_1(t) - X_2(t) = 0] = 1.$$

Existence: We will prove the existence result by constructing a convergent sequence of approximations to equation starting with $X_0(t) = X(0)$. The iterative scheme is as follows:

$$X_{i+1}(t) = X(0) + \int_0^t b(s, X_i(s))ds + \int_0^t \sigma(s, X_i(s))dB(s), \tag{7.12}$$

and using estimates similar to those obtained in the above uniqueness part of the proof, we obtain the inequality

$$\mathbf{E}[X_{i+1}(t) - X_i(t)]^2 \tag{7.13}$$

$$= \mathbf{E}\left[\int_0^t [b(s, X_i(t)) - b(s, X_{i-1}(t))]ds + \int_0^t [\sigma(s, X_i(t)) - \sigma(s, X_{i-1}(t))]dB(s)\right]^2$$

$$\leq L \int_0^t \mathbf{E}|X_i(t) - X_{i-1}(t)|^2 ds,$$

Iterating this scheme n times, we obtain the inequality

$$\mathbf{E}[X_{n+1}(t) - X_n(t)]^2 \leq L \int_0^t \frac{(t-s)^{n-1}}{(n-1)!} \mathbf{E}|X_1(t) - X_0(t)|^2 ds. \tag{7.14}$$

On the other hand

$$\mathbf{E}|X_1(t) - X_0(t)|^2 = \mathbf{E}\left[\int_0^t b(s, X(0))\, ds\right]^2 + \mathbf{E}\int_0^t \sigma^2(s, X(0))\, ds$$

$$\leq LTK^2(1 + \mathbf{E}\eta^2(0)).$$

which implies the existence of a constant C such that

$$\mathbf{E}[X_{n+1}(t) - X_n(t)]^2 \leq C\frac{L^n T^n}{n!}. \tag{7.15}$$

Since,

$$\sup_{t \in [0,T]} |X_{n+1}(t) - X_n(t)| \leq \int_0^t |b(s, X_n(t)) - b(s, X_{n-1}(t))| ds$$

$$+ \sup_{t \in [0,T]} \left|\int_0^t [\sigma(s, X_{n+1}(t)) - \sigma(s, X_n(t))]dB(s)\right|$$

and in view of the inequality

$$\mathbf{E}\sup_{t \in [0,T]} \left|\int_0^t X(s)dB(s)\right|^2 \leq 4\int_0^T \mathbf{E}X^2(s)ds$$

implied by the maximal inequality for stochastic integrals (see, ...) , we obtain that

$$\mathbf{E}\sup_{t \in [0,T]} |X_{n+1}(t) - X_n(t)|2T \int_0^T \mathbf{E}K^2|X_n(s) - X_{n-1}(s)|^2 ds$$

$$+ 8K^2 \int_0^T \mathbf{E}|X_n(s) - X_{n-1}(s)|^2 ds \leq \frac{C_1 L^{n-1} T^{n-1}}{(n-1)!},$$

with $C_1 = K^2(2T + 8)$. Consequently, the series

$$X(0) + \sum_{n=0}^{\infty} [X_{n+1}(t) - X_n(t)] = \lim_{n \to \infty} X_n(t) =: X(t) \tag{7.16}$$

converges uniformly on $[0, T]$ with probability 1, because

$$\sum_{n=0}^{\infty} \mathbf{P}\left[\sup_{t\in[0,T]} |X_{n+1}(t) - X_n(t)| > \frac{1}{n^2}\right] \leq \sum_{n=0}^{\infty} \frac{C_1 L^{n-1} T^{n-1}}{(n-1)!} n^4 < \infty.$$

Thus, the limit gives the solution of Equation (7.6), which is almost surely continuous and adapted to the sigma-algebra generated by $B(t), t \in [0, T]$..

Property (ii) can be demonstrated as follows. First, observe that,

$$\mathbf{E}X_n^2(t) \leq 3\left\{\mathbf{E}X^2(0) + \mathbf{E}\left[\int_0^t b(s, X_{n-1}(s)ds\right]^2 + \mathbf{E}\left[\int_0^t \sigma(s, X_{n-1}(s)ds\right]^2\right\}$$

$$\leq 3\mathbf{E}X(0) + 3L \int_0^t \mathbf{E}X_{n-1}^2(s)ds.$$

Iterating this inequality, we obtain

$$\mathbf{E}X_n^2(t) \leq 3EX^2(0) + 3EX^2(0)3Lt + (3L)^2\mathbf{E}\int_0^t (t-s)\mathbf{E}X_{n-2}^2(s)ds$$

$$\leq 3\mathbf{E}X^2(0) + 3\mathbf{E}X^2(0)3Lt + 3\mathbf{E}X^2(0)\frac{(3Lt)^2}{2} + \cdots \leq 3\mathbf{E}X^2(0)e^{3Lt}.$$

which, passing to the limit, $n \to \infty$, gives Property (ii), and completes the proof of the Theorem. □

7.3 SDEs WITH COEFFICIENTS DEPENDING ONLY ON TIME

In this case, Equation (7.2) takes the form

$$dX(t) = b(t)\, dt + \sigma(t)\, dB(t), \qquad (7.17)$$

and the solution, obviously, is of the form

$$X(t) = X(0) + \int_0^t b(s)\, ds + \int \sigma(s)\, dB(s).$$

It has independent Gaussian increments, and

$$\mathbf{E}(X(t) - X(0)) = \int_0^t b(s)\, ds, \quad \text{and} \quad \text{Var}(X(t) - X(0)) = \int_0^t \sigma^2(s)\, ds.$$

There are other SDEs that can be reduced to form (7.6) by a change of variables

$$Y(t) = g(t, X(t))$$

Indeed, in view of (7.4–7.5), we have to have

$$\hat{b}(t) = \frac{\partial}{\partial t}g(t, x) + b(t, x)\frac{\partial}{\partial x}g(t, x) + \frac{1}{2}\sigma^2(t, x)\frac{\partial^2}{\partial^2 x}g(t, x) \qquad (7.18)$$

and

$$\hat{\sigma}(t) = \sigma(t, x) \frac{\partial}{\partial x} g(t, x).$$ (7.19)

From (7.11) we get

$$\frac{\partial^2}{\partial t \partial x} g(t, x) = \frac{\hat{\sigma}'_t(t)\sigma(t, x) - \hat{\sigma}(t)\sigma'_t(t, x)}{\sigma^2)t, x)}, \qquad \frac{\partial^2}{\partial^2 x} g(t, x) = -\frac{\hat{\sigma}(t)\sigma'_x(t, x)}{\sigma^2)t, x)}$$ (7.20)

so that differentiating (7.11) with respect to x and substituting (7.12) we get the condition

$$\frac{\hat{\sigma}'_t(t)}{\hat{\sigma}(t)} = \sigma(t, x) \left(\frac{\sigma'_t(t, x)}{\sigma^2(t, x)} - \frac{\partial}{\partial x} \left(\frac{b(t, x)}{\sigma(t, x)} \right) + \frac{1}{2} \frac{\partial^2}{\partial x^2} \sigma(t, x) \right)$$ (7.21)

which, after taking another derivative with respect to x, implies the condition

$$\frac{\partial}{\partial x} \left[\sigma(t, x) \left(\frac{\sigma'_t(t, x)}{\sigma^2(t, x)} - \frac{\partial}{\partial x} \left(\frac{b(t, x)}{\sigma(t, x)} \right) + \frac{1}{2} \frac{\partial^2}{\partial x^2} \sigma(t, x) \right) \right] = 0$$ (7.22)

(1) Explicit solutions?

(2) Properties of solutions.

7.4 POPULATION GROWTH MODEL AND OTHER EXAMPLES

7.4.1 Population Growth Model

Random population X_t grows according to the dynamics

$$\frac{dX_t}{dt} = b_t X_t,$$

with the growth rate

$$b_t = r + \beta \cdot \dot{B}_t, \qquad r, \beta > 0.$$

The model can be rewritten as a stochastic differential equation

$$dX_t = rX_t \, dt + \beta X_t \, dB_t.$$ (7.23)

To see how we can apply Itô's formula let us reformulate it as

$$\frac{dX_t}{X_t} = r \, dt + \beta \, dB_t,$$ (7.24)

which suggests an exploration of the $dg(t, X_t)$ with $g(t, x) = \log x$. Indeed,

$$d(\log X_t) = \frac{1}{X_t} dX_t + \frac{1}{2} \left(-\frac{1}{X_t^2} \right) (dX_t)^2 = \frac{dX_t}{X_t} - \frac{1}{2} \beta^2 \, dt.$$

In other words,

$$\frac{dX_t}{X_t} = d(\log X_t) + \frac{1}{2} \beta^2 \, dt,$$

and substituting (7.7), we get

$$r\,dt + \beta\,dB_t = d(\log X_t) + \frac{1}{2}\beta^2\,dt,$$

which gives the solution of (7.6):

$$X_t = X_0 \cdot e^{\left(r - \beta^2/2\right)t + \beta B_t} \tag{7.25}$$

Since $B_t/t \to 0$, as $t \to \infty$, we have, a.s.

$$\lim_{t \to \infty} X_t = \infty, \qquad \text{for} \qquad r > \beta^2/2,$$

and

$$\lim_{t \to \infty} X_t = 0, \qquad \text{for} \qquad r < \beta^2/2,$$

Surprisingly, for any $\beta > 0$,

$$\lim_{t \to \infty} \mathbf{E}X_t = \infty. \tag{7.26}$$

Indeed, applying Itô's formula to the exponential function we get

$$de^{\beta B_t} = \beta e^{\beta B_t}\,dB_t + \frac{1}{2}\beta^2 e^{\beta B_t}\,dt.$$

Thus

$$\mathbf{E}e^{\beta B_t} = \mathbf{E}\int_0^t \beta e^{\beta B_s}\,dB_s + \frac{1}{2}\int_0^t \beta^2 \mathbf{E}e^{\beta B_s}\,ds = \frac{1}{2}\beta^2 \int_0^t \mathbf{E}e^{\beta B_s}\,ds.$$

and, necessarily,

$$\mathbf{E}e^{\beta B_t} = e^{t\beta^2/2}.$$

so that

$$\mathbf{E}X_t = e^{rt}.$$

7.4.2 Ornstein–Uhlenbeck Process

The Ornstein–Uhlenbeck process is defined by the stochastic differential equation of the form

$$dX_t = bX_t dt + \sigma dB_t,$$

where b, and σ are constants. The equation can be solved in a way similar to the approach taken above for the population growth model, by multiplying both sides by the integrating factor e^{-t} and applying Itô's formula to the differential $d(e^{-t}X_t)$.

7.5 SYSTEMS OF SDEs AND VECTOR-VALUED ITÔ'S FORMULA

In this section, we briefly review the theory of systems of stochastic differential equations

$$dX_1(t) = b_1(t, X_1(t), \dots, X_n(t)) + \sum_{j=1}^m \sigma_{ij}(t, X_1(t), \dots, X_n(t))B_j(t),$$

$$\dots\dots\dots \tag{7.27}$$

$$dX_n(t) = b_n(t, X_1(t), \ldots, X_n(t)) + \sum_{j=1}^{m} \sigma_{nj}(t, X_1(t), \ldots, X_n(t)) B_j(t),$$

where $B_1(t), \ldots, B_m(t)$, are independent standard Brownian motions. The system can be written in a more compact vector/matrix form,

$$d\vec{X}(t) = \vec{b}(t, \vec{X}(t))\,dt + \bar{\bar{\sigma}}(t, \vec{X}(t))\,d\vec{B}(t),$$

where

$$\vec{B}(t) = (B_1(t), \ldots, B_m(t))^T,$$
$$\vec{X}(t) = (X_1(t), \ldots, X_n(t))^T,$$
$$\vec{b}(t, \vec{X}(t)) = \left(b_1(t, \vec{X}(t)), \ldots, b_n(t, \vec{X}(t))\right)^T,$$

and the matrix

$$\bar{\bar{\sigma}}(t, \vec{X}(t)) = \left[\sigma_{ij}(t, \vec{X}(t))\right], \quad i = 1, \ldots, n, \quad j = 1, \ldots, m.$$

In the vector context Itô's formula takes the following form, with the proof similar to the proof of the one-dimensional case (Theorem 6.2) :

Theorem 7.2. *Let the function*

$$\vec{g} : [0, \infty) \times \mathbf{R}^n \ni (t, x_1, \ldots, x_n) \mapsto (g_1, \ldots g_m) \in \mathbf{R}^m,$$

be continuously twice differentiable. If $\vec{X}(t)$ is a vector-valued Itô's integral process, then the process

$$\vec{Y}(t) = \vec{g}(t, \vec{X}(t))$$

is again a vector-valued Itô's integral process with the differential components

$$d_k(t) = \frac{\partial g_k}{\partial t}(t, \vec{X}(t))dt + \sum_{i=1}^{n} \frac{\partial g_k}{\partial x_i}(t, \vec{X}(t))\,dX_i + \frac{1}{2}\sum_{i,j=1}^{n} \frac{\partial^2 g_k}{\partial x_i \partial x_j}(t, \vec{X}(t))\,dX_i\,dX_j$$

$k = 1, \ldots m$, with the stochastic multiplication rule,

$$dB_i\,dB_j = \delta_{ij}\,dt, \quad dB_i\,dt = 0, \quad i, j = 1, \ldots, n,$$

where δ_{ij} is the Kronecker delta equal to 0, if $i \neq j$, and 1, if $i = j$.

Example 7.7 (Electrical circuits with noise). *Let us return to the problem of electrical circuit with noise that was informally introduced in Section 7.1. The current intensity $i(t)$ of an RLC circuit (in series) is described by the SDE of the second order,*

$$Li''(t) + Ri'(t) + \frac{1}{C}i(t) = V'(t) + \sigma B(t) \tag{7.28}$$

where $V(t)$ is the external voltage and the Brownian motion, $B(t)$, represents the noise. The above equation can be rewritten as a first-order vector SDE in two dimensions, by introducing the notation,

$$X_1(t) = i(t), \qquad X_2(t) = i'(t), \quad \vec{X}(t) = (X_1(t), X_2(t))^T. \qquad (7.29)$$

Now (7.28), can be put in the vector form,

$$\vec{X}'(t) = \begin{bmatrix} X_1'(t) \\ X_2'(t) \end{bmatrix} = \begin{bmatrix} X_2'(t) \\ -\frac{R}{L}X_2(t) - \frac{1}{LC}X_1(t) + \frac{1}{L}V'(t) + \frac{\sigma}{L}B(t) \end{bmatrix}$$

$$= \begin{bmatrix} 1 \\ -\frac{R}{L} \end{bmatrix} X_2(t) + \begin{bmatrix} 0 \\ -\frac{1}{LC} \end{bmatrix} X_1(t) + \begin{bmatrix} 0 \\ -\frac{V'(t)}{L} \end{bmatrix} + \begin{bmatrix} 0 \\ -\frac{\sigma}{L} \end{bmatrix} B(t).$$

Formally, as an SDE, (7.28) now takes the form

$$d\vec{X}(t) = \bar{\bar{a}}\vec{X}(t)\,dt + \vec{h}\,dt + \vec{k}\,dB(t) \qquad (7.30)$$

where

$$\bar{\bar{a}} = \begin{bmatrix} 0 & 1 \\ -\frac{1}{LC} & -\frac{R}{L} \end{bmatrix}. \qquad \vec{h}(t) = \begin{bmatrix} 0 \\ -\frac{V'(t)}{L} \end{bmatrix}, \qquad \vec{k} = \begin{bmatrix} 0 \\ -\frac{\sigma}{L} \end{bmatrix}.$$

An application of the vector Itô's formula to the function

$$\vec{g}(t, \vec{x}) = \exp(-\bar{\bar{a}}t)\vec{x},$$

gives us the differential

$$d\big(\exp(-\bar{\bar{a}}t)\vec{X}(t)\big) = (-\bar{\bar{a}})\exp(-\bar{\bar{a}}t)\vec{X}(t)\,dt + \exp(-\bar{\bar{a}}t)\,dX(t). \qquad (7.31)$$

Multiplying both sides of SDE (7.30) by $\exp(-\bar{\bar{a}}t)$ we get

$$\exp(-\bar{\bar{a}}t)d\vec{X}(t) = \exp(-\bar{\bar{a}}t)\bar{\bar{a}}\vec{X}(t)\,dt + \exp(-\bar{\bar{a}}t)\vec{h}(t)\,dt + \exp(-\bar{\bar{a}}t)\vec{k}\,dB(t)$$

which together with (7.31) yields the equation

$$\exp(-\bar{\bar{a}}t)d\vec{X}(t) = \vec{X}(0) + \int_0^t \exp(-\bar{\bar{a}}s)\vec{h}(s)\,ds + \int_0^t \exp(-\bar{\bar{a}}s)\vec{k}\,dB(s).$$

Finally, we obtain the solution,

$$d\vec{X}(t) = \exp(\bar{\bar{a}}t)\left[\vec{X}(0) + \int_0^t \exp(-\bar{\bar{a}}s)\big(\vec{h}(s) + \bar{\bar{a}}\vec{k}B(s)\big)\,ds + \exp(-\bar{\bar{a}}t)\vec{k}B(t)\right].$$

Example 7.8 (Bessel processes). *The Bessel process $R(t)$ is a radial processes created from two independent Brownian motion components, $\vec{B}(t) = (B_1(t), B_2(t))$, through the formula*

$$R(t) = \big(B_1^2(t) + B_2^2(t)\big)^{1/2}. \qquad (7.32)$$

An application of the two-dimensional Itô's formula gives the stochastic differential

$$dR(t) = \big(B_1^2(t) + B_2^2(t)\big)^{-1/2}\left(B_1(t)dB_1(t) + B_2(t)dB_2(t) + \frac{1}{2}dt\right).$$

Example 7.9 (Brownian Motion on the unit circle). *Consider a function*

$$\vec{g}(t, x) = (\cos x, \sin x) \in \mathbf{R}^2,$$

defined for all $x \in R$ and taking values in the unit circle in the plane, and define the process

$$\vec{Y}(t) = \vec{g}(t, B(t)) = (\cos B(t), \sin B(t)).$$

The process is, again, a stochastic integral with

$$dY_1(t) = \sin B(t)\, dB(t) - \frac{1}{2}\cos B(t)\, dt,$$

$$dY_2(t) = \cos B(t)\, dB(t) - \frac{1}{2}\sin B(t)\, dt.$$

Note that $\vec{Y}(t)$ solves the vector SDE,

$$d\vec{Y}(t) = -\frac{1}{2}\vec{Y}(t)\, dt + \bar{\bar{a}}\vec{Y}(t)\, dB(t),$$

with the matrix,

$$\bar{\bar{a}} = \begin{bmatrix} 0 & -1 \\ 1 & 0 \end{bmatrix}.$$

7.6   NUMERICAL SOLUTION OF STOCHASTIC DIFFERENTIAL EQUATIONS

The simplest scheme for the numerical solution of the stochastic differential equation

$$X_t = X_0 \int_0^t b(s, X_s)ds + \int_0^t \sigma(s, X)s)\, dB_s$$

is analogous to the usual time-discrete Euler scheme for deterministic differential equations. In the stochastic case, it is usually called the *Euler-Maruyama* scheme, and it has the iterative form

$$Y(t_{n+1}) = Y(t_n) + b(t_n, Y(t_n))\Delta t_n + \sigma(t_n, Y(t_n))\Delta B(t_{n+1}). \tag{7.33}$$

where

$$\Delta t_n = \Delta t = \frac{T}{N}, \quad \text{and} \quad \Delta B(t_n) \quad \text{are } i.i.d. \ N(0, \Delta t)$$

The accuracy of the Euler-Maruyama scheme is not particularly good. Namely, the convergence is of order square root of the time step Δt, and the method works best if the coefficients b, and σ, do not vary too much. More precisely, we have the following

Theorem 7.3. *Assume that $\mathbf{E}|X_0|^2 < \infty$, that the SDE's coefficients satisfy the inequalities*

$$|b(t, x) - b(t, y)| + |\sigma(t, x) - \sigma(t, y)| \le K_1|x - y|, \tag{7.34}$$

$$|b(t,x)| + |\sigma(t,x)| \leq K_2(1 + |x|), \tag{7.35}$$

and

$$|b(s,x) - b(t,x)| + |\sigma(s,x) - \sigma(t,x)| \leq K_3(1 + |x|)|s - t|^{1/2}, \tag{7.36}$$

for some constants, K_1, K_2, K_3, all $s, t \in [0,T]$, and any $x, t \in \mathbf{R}$. Additionally, assume that the initial condition $Y_0^{\Delta t}$ of the numerical scheme satisfies the condition

$$\mathbf{E}\left(|X_0 - Y_0^{\Delta t}|^2\right)^{1/2} \leq L\Delta t^{1/2}, \tag{7.37}$$

for some constant L independent of Δt. Then, there exists a constant K, independent of Δt, such that

$$\mathbf{E}(|X_T - Y^{\Delta t}(T)|) \leq K\Delta t^{1/2}. \tag{7.38}$$

Proof. $\cdots$ □

A more accurate numerical method is provided by the following *Millstein scheme*, which uses the second-order Taylor approximation motivated by the Itô's formula:

$$Y(t_{n+1}) = Y(t_n) + b(t_n, Y(t_n))\Delta t_n + \sigma(t_n, Y(t_n))\Delta B(t_{n+1})$$
$$+ \frac{1}{2}\sigma(t_n, Y(t_n))\frac{\partial}{\partial x}\sigma(t_n, Y(t_n))\left[(\Delta B(t_{n+1}))^2 - \Delta t\right] \tag{7.39}$$

Under conditions similar to those listed in Theorem 7.2, one can prove that the Milstein scheme is of order 1, that is

$$\mathbf{E}(|X_T - Y^{\Delta t}(T)|) \leq K\Delta t, \tag{7.40}$$

for some constant K which is independent of Δt.[2]

Do Feynman-Kac ???

---

[2]For more details, see P.E. Kloeden and E. Platen, *Numerical Solution of Stochastic Differential Equations,* 632 pp., Springer-Verlag, Berlin-Heidelberg-New York, 1992.

CHAPTER 8

Asymmetric Exclusion Processes and Their Scaling Limits

8.1 ASYMMETRIC EXCLUSION PRINCIPLES

Particles occupy integer lattice sites on the real line. Description: $X(k, t) = 1$, if site k is occupied at time t, and 0 otherwise. They obey the exclusion principle: two particles cannot occupy the same site at the same time.

This is the example considered by Kipnis (1986), Benassi and Fouque (1987), and Srinivasan (1991,1993). Another starting point is the observation that the queuing system consisting of an infinite series of queues can be interpreted in the language of the one-dimensional nearest neighbor simple exclusion process (see, e.g., Liggett, 1985). Indeed, if the lattice location of the i-th particle is denoted by x_i then, in view of the exclusion dynamics and nearest neighbor jumps, at time t

$$\cdots < x_{-1}(t) < x_0(t) < x_1(t) < \ldots \tag{8.1}$$

Assume that the rate of this process is 1. If we denote by $\eta_i(t)$, the random variable equal to the number of empty sites between $x_i(t)$ and $x_{i+1}(t)$, then $\eta_i(t)$ can be considered as the length of the ith queue for an infinite queuing system with single servers in series, each with an exponential service time with intensity 1. Indeed, when the ith particle jumps to the right by one unit, then $\eta_i(t)$ changes into $\eta_i(t) - 1$ which means that the service for one customer was completed at the ith server, and η_{i+1} is changed to $\eta_{i+i}(t) + 1$, which means that a new customer was added to the queue at the $(i+1)$st server. In other words, the customer in the ith queue is served in exponential time with rate 1 and then joins the $(i - 1)$st queue with probability p and $(i + 1)$st queue with probability $1 - p$.

Another way to code the asymmetric exclusion interacting particle system is by listing its states

$$X(t) = \{X(k, t) : t \geq 0, k \in \mathbf{Z}\} \in \{0, 1\}^{\mathbf{Z}}, \tag{8.2}$$

DOI: 10.1201/9781003216759-8

The set $\{k : X(k, t) = 1\} \subset \mathbf{Z}$ is the set of occupied sites at time t. In the totally asymmetric case $p = 1$, the infinitesimal generator for the Markov process $X(t)$ (which does exist, see, e.g., Liggett, 1985)

$$\mathcal{L}f(X) = \sum_{k \in Z} X(k)\big(1 - X(k+1)\big)\big[f(X^{k,k+1}) - f(X)\big], \tag{8.3}$$

where the state $X^{k,k+1}$ is obtained from the state X by setting $X(k) = 0, X(k+1) = 1$ and keeping the other values fixed.

The above system's dynamics can also be encoded in the infinite system of ordinary stochastic differential equations

$$\begin{aligned}
dX(t, k) = {}& X(t^-, k-1)\big[1 - X(t^-, k)\big]\, dP(t, k-1) \\
& - X(t^-, k)\big[1 - X(t^-, k+1)\big]\, dP(t, k) \\
& + X(t^-, k+1)\big[1 - X(t^-, k)\big]\, dQ(t, k+1) \\
& - X(t^-, k)\big[1 - X(t^-, k-1)\big]\, dQ(t, k) \tag{8.4}
\end{aligned}$$

where $P(t, k)$ and $Q(t, k)$, $k \in \mathbf{Z}$, are independent Poisson processes with intensities p and $(1 - p)$, representing jumps to the right and jumps to the left, respectively.

8.2   SCALING LIMIT

Define the hyperbolic rescalings

$$X^h(t, x) = \sum_{k \in \mathbf{Z}} X\left(\frac{t}{h}, k\right) \mathbf{1}_{[hk, h(k+1))}(x), \tag{8.5}$$

$$P^h(t, x) = \sum_{k \in \mathbf{Z}} P\left(\frac{t}{h}, k\right) \mathbf{1}_{[hk, h(k+1))}(x), \tag{8.6}$$

$$Q^h(t, x) = \sum_{k \in \mathbf{Z}} Q\left(\frac{t}{h}, k\right) \mathbf{1}_{[hk, h(k+1))}(x), \tag{8.7}$$

and introduce notation

$$F_{\pm h}u(x) = u(x)\big(1 - u(x \pm h)\big), \tag{8.8}$$

$$D_{\pm h}u(x) = \pm\frac{u(x \pm h) - u(x)}{h}. \tag{8.9}$$

A direct verification shows that the system (8.4) can now be written in the form

$$\begin{aligned}
dX^h(t, x) = {}& -D_{-h}\big[F_h\big(X^h(t^-, x)\big)\, d(hP^h(t, x))\big] \\
& + D_h\big[F_{-h}\big(X^h(t^-, x)\big)\, d(hQ^h(t, x))\big]. \tag{8.10}
\end{aligned}$$

Theorem 8.1 (Benassi and Fouque, 1987). *Let $p \neq 1/2$. As $h \to 0$, the solution $X^h(t, x)\, dx$ of (8.10) converges weakly to $u(t, x)\, dx$, where $u(t, x)$ is a decreasing and right continuous in the x-variable weak solution of the nonlinear Cauchy* *problem*

$$\frac{\partial u}{\partial t} + (2p - 1)\frac{\partial B(u)}{\partial x} = 0, \tag{8.11}$$

$$u(0, x) = u_0(x) = b\mathbf{1}_{(-\infty, 0]}(x) + a\mathbf{1}_{(0, \infty]}(x), \tag{8.12}$$

with some $0 \leq a < b < \infty$ *and*

$$B(u) = u(1 - u). \tag{8.13}$$

Moreover, for all t, x, *we have* $a \leq u(t, x) \leq b$.

Recall that the weak solution is understood in the following sense: for every smooth function $\phi : \mathbf{R}^+ \times \mathbf{R} \to \mathbf{R}$ with compact support,

$$\int_{\mathbf{R}^+} \int_{\mathbf{R}} [u\phi_t + (2p - 1)F(u)\phi_x] \, dx \, dt = - \int_{\mathbf{R}} u_0(x)\phi(0, x) \, dx. \tag{8.14}$$

Heuristically, the result is plausible, since, as $h \to 0$ in (8.10), $hP^h(t, x) \to pt$, $hQ^h(t, x) \to (1 - p)t$, $D_{\pm h} \to \partial/\partial x$, and $F_{\pm h} \to F$.

8.3 OTHER QUEUING REGIMES RELATED TO NON-NEAREST NEIGHBOR SYSTEMS

In this section, we describe an extension of the queues-in-series/nearest-neighbor-asymmetric-exclusion-system model discussed in Section 8.2 in the sense that it will permit longer but still finite-range interactions.[1]

Consider an interacting particle system $X(t) = \{X(k, t) : t \geq 0, k \in \mathbf{Z}\} \in \{0, 1\}^{\mathbf{Z}}$, operating under the following dynamics:

(i) There is always at most one particle per site;

(ii) Each particle has an exponential alarm clock with rate 1, the clocks are independent of each other. When the particle's clock rings, the particle moves to the right (left) with probability p $(1 - p)$ by one step but only if the site it is moving to is unoccupied *and* the site immediately to the right (left) of the target site is also unoccupied.

In the totally asymmetric case $p = 1$, the model has an infinitesimal generator of the form

$$\mathcal{L}f(X) = \sum_{k \in Z} X(k)\big(1 - X(k + 1)\big)\big(1 - X(k + 2)\big)\big[f(X^{k,k+1}) - f(X)\big], \tag{8.15}$$

As before, we can interpret the above particle system as a queuing network of server in series with the number of empty sites between the particles i and $(i + 1)$

---

[1]See, also, Margolius and Woyczynski, ...

(i.e., the interparticle distance minus one) being interpreted as the queue length at the ith server, but the regime we are now considering will activate the server only if the queue length is at least 2. In other words, the mental picture could be that of the network of secretaries in series, with each secretary gossiping with the first customer until another customer shows up and only then beginning to serve the first customer in exponential time.

In more generality, suppose the secretaries will not attend to a customer until there is a queue of m customers waiting. The system's dynamics can then be encoded in the infinite system of ordinary stochastic differential equations

$$
\begin{aligned}
dX(t,k) \\
= X(t^-,k-1)\big[1-X(t^-,k)\big] \cdots \cdot \big[1-X(t^-,k+m-1)\big]\,dP(t,k-1) \\
- X(t^-,k)\big[1-X(t^-,k+1)\big] \cdots \cdot \big[1-X(t^-,k+m)\big]\,dP(t,k) \\
+ X(t^-,k+1)\big[1-X(t^-,k)\big] \cdots \cdot \big[1-X(t^-,k-m+1)\big]\,dQ(t,k+1) \\
- X(t^-,k)\big[1-X(t^-,k-1)\big] \cdots \cdot \big[1-X(t^-,k-m)\big]\,dQ(t,k), \qquad (8.16)
\end{aligned}
$$

where $P(t,k)$ and $Q(t,k)$, $k \in \mathbf{Z}$, are independent Poisson processes with intensities p and $(1-p)$, representing jumps to the right and jumps to the left, respectively.

Define the hyperbolic rescalings $X^h(t,x)$, $P^h(t,x)$, and $Q^h(t,x)$ as in (8.5–8.7) and the difference operator $D_{\pm h}u(x) = $ as in (8.9), but,

$$
F_{\pm h}u(x) = u(x)\big(1-u(x\pm h)\big) \cdots \cdot \big(1-u(x\pm mh)\big), \qquad (8.17)
$$

In this notation, one can verify directly that X_h satisfies the equation

$$
\begin{aligned}
dX^h(t,x) = -D_{-h}\big[F_h(X^h(t^-,x))\,d(hP^h(t,x))\big] \\
+ D_h\big[F_{-h}(X^h(t^-,x))\,d(hQ^h(t,x))\big].
\end{aligned} \qquad (8.18)
$$

Indeed, working backward, beginning with (8.18) and substituting for $D_{\pm h}$, we obtain

$$
\begin{aligned}
dX^h(t,x) = \frac{1}{h}\Big[F_h(X^h(t^-,x-h))d(hP^h(t,x-h)) \\
- F_h(X^h(t^-,x))d(hP^h(t,x))\Big] \\
+ \frac{1}{h}\Big[F_{-h}(X^h(t^-,x+h))d(hQ^h(t,x+h)) \\
- F_{-h}(X^h(t^-,x))d(hQ^h(t,x))\Big].
\end{aligned} \qquad (8.19)
$$

Next, substituting (8.17) for $F_{\pm h}$, gives

$$
\begin{aligned}
dX^h(t,x) = +\frac{1}{h}\Big[X^h(t^-,x-h)\big[1-X^h(t^-,x)\big] \cdots \cdot \big[1-X^h(t^-,x+h(m-1))\big] \\
\times\, d(hP^h(t,x-h)) \\
- X^h(t^-,x)\big[1-X^h(t^-,x+h)\big] \cdots \cdot \big[1-X^h(t^-,x+mh)\big]
\end{aligned}
$$

$$\times d(hP^h(t,x))\Big]$$

$$+ \frac{1}{h}\Big[X^h(t^-,x+h)\big[1-X^h(t^-,x)\big]\cdot\cdots\cdot\big[1-X^h(t^-,x-(m-1)h)\big]$$

$$\times d(hQ^h(t,x+h))$$

$$- X^h(t^-,x)\big[1-X^h(t^-,x-h)\big]\cdot\cdots\cdot\big[1-X^h(t^-,x-mh)\big]$$

$$\times d(hQ^h(t,x))\Big]. \tag{8.20}$$

From the definition of $P^h(t,x)$, and $Q^h(t,x)$, we have

$$dX^h(t,x) = +\frac{1}{h}\Big[X^h(t^-,x-h)\big[1-X^h(t^-,x)\big]\cdot\cdots\cdot\big[1-X^h(t^-,x+h(m-1))\big]$$

$$\times d\Big(h\sum_{k\in\mathbf{Z}}P\Big(\frac{t}{h},k\Big)\mathbf{1}_{[hk,h(k+1))}(x-h)\Big)$$

$$-X^h(t^-,x)\big[1-X^h(t^-,x+h)\big]\cdot\cdots\cdot\big[1-X^h(t^-,x+mh)\big]$$

$$\times d\Big(h\sum_{k\in\mathbf{Z}}P\Big(\frac{t}{h},k\Big)\mathbf{1}_{[hk,h(k+1))}(x)\Big)\Big]$$

$$+ \frac{1}{h}\Big[X^h(t^-,x+h)\big[1-X^h(t^-,x)\big]\cdot\cdots\cdot\big[1-X^h(t^-,x-(m-1)h)\big]$$

$$\times d\Big(h\sum_{k\in\mathbf{Z}}Q\Big(\frac{t}{h},k\Big)\mathbf{1}_{[hk,h(k+1))}(x+h)\Big)$$

$$-X^h(t^-,x)\big[1-X^h(t^-,x-h)\big]\cdot\cdots\cdot\big[1-X^h(t^-,x-mh)\big]$$

$$\times d\Big(h\sum_{k\in\mathbf{Z}}Q\Big(\frac{t}{h},k\Big)\mathbf{1}_{[hk,h(k+1))}(x)\Big)\Big]. \tag{8.21}$$

The definition of $X^h(t,x)$ and substitution $x=hj$ gives

$$dX^h(t,hj) = +\frac{1}{h}\Big[X\Big(\frac{t}{h},j-1\Big)\Big(1-X\Big(\frac{t}{h},j\Big)\Big)\cdot\cdots\cdot\Big(1-X\Big(\frac{t}{h},j+m-1\Big)\Big)$$

$$\times d\Big(hP\Big(\frac{t}{h},j-1\Big)\Big)$$

$$- X\Big(\frac{t}{h},j\Big)\Big(1-X\Big(\frac{t}{h},j+1\Big)\Big)\cdot\cdots\cdot\Big(1-X\Big(\frac{t}{h},j+m\Big)\Big)$$

$$\times d\Big(hP\Big(\frac{t}{h},j\Big)\Big)\Big]$$

$$+ \frac{1}{h}\Big[X\Big(\frac{t}{h},j+1\Big)\Big(1-X\Big(\frac{t}{h},j\Big)\Big)\cdot\cdots\cdot\Big(1-X\Big(\frac{t}{h},j-(m-1)\Big)\Big)$$

$$\times d\Big(hQ\Big(\frac{t}{h},j+1\Big)\Big)$$

$$
\begin{aligned}
&- X\!\left(\frac{t}{h}, j\right)\left(1 - X\!\left(\frac{t}{h}, j-1\right)\right)\cdots\cdots\left(1 - X\!\left(\frac{t}{h}, j-m\right)\right)\\
&\times d\!\left(hQ\!\left(\frac{t}{h}, j\right)\right)\Bigg],
\end{aligned}
\tag{8.22}
$$

so that, finally,

$$
\begin{aligned}
dX\!\left(\frac{t}{h}, j\right) &= \Bigg[X\!\left(\frac{t}{h}, j-1\right)\left(1 - X\!\left(\frac{t}{h}, j\right)\right)\cdots\left(1 - X\!\left(\frac{t}{h}, j+m-1\right)\right)\\
&\quad \times d\!\left(P\!\left(\frac{t}{h}, j-1\right)\right)\\
&\quad - X\!\left(\frac{t}{h}, j\right)\left(1 - X\!\left(\frac{t}{h}, j+1\right)\right)\cdots\left(1 - X\!\left(\frac{t}{h}, j+m\right)\right)\\
&\quad \times d\!\left(P\!\left(\frac{t}{h}, j\right)\right)\Bigg]\\
&\quad + \Bigg[X\!\left(\frac{t}{h}, j+1\right)\left(1 - X\!\left(\frac{t}{h}, j\right)\right)\cdots\left(1 - X\!\left(\frac{t}{h}, j-(m-1)\right)\right)\\
&\quad \times d\!\left(Q\!\left(\frac{t}{h}, j+1\right)\right)\\
&\quad - X\!\left(\frac{t}{h}, j\right)\left(1 - X\!\left(\frac{t}{h}, j-1\right)\right)\cdots\left(1 - X\!\left(\frac{t}{h}, j-m\right)\right)\\
&\quad \times d\!\left(Q\!\left(\frac{t}{h}, j\right)\right)\Bigg],
\end{aligned}
\tag{8.23}
$$

which justifies (8.19).

Now note that in view of (8.19), heuristically, it is clear that

$$
\begin{aligned}
\lim_{h\to 0} dX^h(t, x) &= -\frac{\partial}{\partial x}\Big[u(t, x)(1 - u(t, x))^m\Big]\, d(pt)\\
&\quad + \frac{\partial}{\partial x}\Big[u(t, x)(1 - u(t, x))^m\Big]\, d((1-p)t),
\end{aligned}
\tag{8.24}
$$

which provides an argument in support of the following result about approximation of density profiles for the GS queuing networks by solutions of a nonlinear hyperbolic equation.

Theorem 8.2. *Let $p \neq 1/2$. As $h \to 0$, the solution $X^h(t, x)\, dx$ of (8.18) converges weakly to $u(t, x)\, dx$, where $u(t, x)$ is a weak solution of the nonlinear Cauchy problem*

$$
\frac{\partial u}{\partial t} + (2p - 1)\frac{\partial F(u)}{\partial x} = 0.
\tag{8.25}
$$

$$
u(0, x) = u_0(x) = b\mathbf{1}_{(-\infty, 0]}(x) + a\mathbf{1}_{(0, \infty)}(x),
\tag{8.26}
$$

with some $0 \leq a < b < \infty$ and

$$
F(u) = [u(t, x)(1 - u(t, x))^m].
\tag{8.27}
$$

Moreover, for all t, x, we have $a \leq u(t, x) \leq b$.

8.4 NETWORKS WITH MULTISERVER NODES AND PARTICLE SYSTEMS WITH STATE-DEPENDENT RATES

The exclusion interacting particle system $X(t) = \{X(t,k) : t \geq 0, k \in \mathbf{Z}\} \in 2^{\mathbf{Z}}$, discussed in this section is operating under the following dynamics:

(i) There is always at most one particle per site;

(ii) Each particle has an exponential alarm clock with a state-dependent rate which also depends on the constants $0 \leq p \leq 1$, and a positive integer N. The rate is p times the minimum of r, the number of empty sites to the right, and N, plus $1 - p$ times the minimum of l, the number of empty sites to the left, and N. When the particle's clock rings, the particle moves to the right by one step with probability

$$p_{r,l} = \frac{p\min\{N, r\}}{p\min\{N, r\} + (1 - p)\min\{N, l\}},$$

but only if the target site to the right is unoccupied. The particle moves to the left with probability

$$1 - p_{r,l} = \frac{(1 - p)\min\{N, l\}}{p\min\{N, r\} + (1 - p)\min\{N, l\}},$$

but only if the target site to the left is unoccupied.

The queuing-theoretic interpretation of this model is analogous to the examples in Sections 8.2–8.3. We have an infinite series of queues. As in the nearest neighbor exclusion system of Section 8.2, we denote by $\eta_i(t)$ the random variable equal to the number of empty sites between $x_i(t)$ and $x_{i+1}(t)$, where $x_i(t)$ is the location of the ith particle, and, as before, $\eta_i(t)$ represents the length of the ith queue. In this model though, we have N identical servers at each particle. When the servers are serving the queue to the right, service for the first customer served is completed at the rate $\min\{N, r\}$. When the servers are serving the queue to the left, service for the first customer served is completed at the rate $\min\{N, l\}$. The former event occurs with probability $p_{r,l}$, and the latter with probability $1 - p_{r,l}$. When the ith particle jumps to the right, $\eta_i(t)$ changes to $\eta_i(t) - 1$ which means that service for one customer was completed at the ith server, and $\eta_{i+1}(t)$ is changed to $\eta_{i+1}(t)+1$, i.e., a new customer is added to the $(i + 1)$st queue. When $p = 1$, customers move through the queueing stations sequentially. When $0 < p < 1$, customers can move back and forth through the queues in the network. When there are fewer in the queue than the number of servers, then this regime corresponds to all customers being in service and the jump occurs when the first of the $\eta_i(t)$ customers has completed service.

The corresponding infinitesimal generator for this regime in the totally asymmetric case is

$$Lf(X) = \sum_{k \in \mathbf{Z}} \left(X(t^-, k) \left[\sum_{m=1}^{N-1} m \left(\prod_{i=1}^{m}[1 - X(t^-, k + i)] \right) X(t^-, k + m + 1) \right. \right.$$

$$+N\left(\prod_{i=1}^{N}[1-X(t^{-},k+i)]\right)\right]\right)\,[f(X^{k,k+1})-f(X)].\qquad(8.28)$$

We will now describe the evolution of this particle system through a system of stochastic differential equations. Note that if the site to the left of site k is occupied with particle i and $\eta_i = 1$, then a jump to the right occurs according to the independent Poisson process $P(t,k)$ with intensity p. This event occurs if $X(t^{-},k-1)(1-X(t^{-},k))X(t^{-},k+1)=1$. More generally, if the site to the left of site k is occupied with particle i and $\eta_i = r < N$, then a jump to the right occurs according to the independent Poisson process $P_r(t,k)$ with intensity rp. This event occurs if $X(t^{-},k-1)(1-X(t^{-},k))(1-X(t^{-},k+1))\cdots(1-X(t^{-},k+r-1))X(t^{-},k+r)=1$. If the site to the left of k is occupied with particle i and $\eta_i \geq N$, then the quantity $X(t^{-},k-1)(1-X(t^{-},k))(1-X(t^{-},k+1))\cdots(1-X(t^{-},k+N))=1$ and a jump to the right is governed by the independent Poisson process $P_N(t,k)$ with intensity Np.

Jumps from the site k are governed by the same rules so that if $X(t^{-},k)(1-X(t^{-},k+1))X(t^{-},k+2)=1$ a jump to the right out of site k occurs according to the independent Poisson process $P(t,k)$ with intensity p. More generally, if site k is occupied with particle i and $\eta_i = r < N$, then a jump to the right occurs according to the independent Poisson process $P_r(t,k)$ with intensity rp. This event occurs if $X(t^{-},k-1)(1-X(t^{-},k))(1-X(t^{-},k+1))\cdots(1-X(t^{-},k+r-1))X(t^{-},k+r)=1$. If the site to the left of k is occupied with particle i and $\eta_i \geq N$, then the quantity $X(t^{-},k-1)(1-X(t^{-},k))(1-X(t^{-},k+1))\cdots(1-X(t^{-},k+N))=1$ and a jump to the right is governed by the independent Poisson process $P_N(t,k)$ with intensity Np. For jumps to the left, we replace $P(t,k)$ with $Q(t,k)$ and p with $(1-p)$ and change signs, but otherwise the analysis is the same.

This analysis leads to the following system ($k \in \mathbf{Z}$) of ordinary stochastic differential equations:

$$dX(t,k) = +X(t^{-},k-1)\left[\sum_{m=1}^{N-1}m\left(\prod_{i=1}^{m}[1-X(t^{-},k+(i-1))]\right)X(t^{-},k+m)\right.$$

$$\left.+N\left(\prod_{i=1}^{N}[1-X(t^{-},k+(i-1))]\right)\right]dP(t,k-1)$$

$$-X(t^{-},k)\left[\sum_{m=1}^{N-1}m\left(\prod_{i=1}^{m}[1-X(t^{-},k+i)]\right)X(t^{-},k+m+1)\right.$$

$$\left.+N\left(\prod_{i=1}^{N}[1-X(t^{-},k+i)]\right)\right]dP(t,k)$$

$$+X(t^{-},k+1)\left[\sum_{m=1}^{N-1}m\left(\prod_{i=1}^{m}[1-X(t^{-},k-(i-1))]\right)X(t^{-},k-m)\right.$$

$$+ N \left(\prod_{i=1}^{N} [1 - X(t^-, k - (i-1))] \right) \Bigg] dQ(t, k+1)$$

$$- X(t^-, k) \Bigg[\sum_{m=1}^{N-1} m \left(\prod_{i=1}^{m} [1 - X(t^-, k-i)] \right) X(t^-, k - m - 1)$$

$$+ N \left(\prod_{i=1}^{N} [1 - X(t^-, k - i)] \right) \Bigg] dQ(t, k), \qquad (8.29)$$

After the hyperbolic rescaling (8.5–8.7), the system can be written as a single equation

$$dX^h(t, x) = - D_{-h} \big[G_h \big(X^h(t^-, x) \big) \, d \big(h P^h(t, x) \big) \big]$$
$$+ D_h \big[G_{-h} \big(X^h(t^-, x) \big) \, d \big(h Q^h(t, x) \big) \big], \qquad (8.30)$$

where

$$G_{\pm h} u(x) = \sum_{m=1}^{N-1} m u(x) \left(\prod_{i=1}^{m} [1 - u(x \pm ih)] \right) u(x \pm (m+1)h)$$

$$+ N u(x) \left(\prod_{i=1}^{N} [1 - u(x \pm ih)] \right), \qquad (8.31)$$

and $D_{\pm h}$ is as defined above in (8.9).

Indeed, substituting for $D_{\pm h}$ and $G_{\pm h}$ in (8.30), we obtain

$$dX^h(t, x) = + \frac{1}{h} \Bigg\{ \Bigg[\sum_{m=1}^{N-1} m X^h(t^-, x - h) \left(\prod_{i=1}^{m} (1 - X^h(t^-, x + (i-1)h)) \right)$$
$$\times X^h(t^-, x + mh)$$
$$+ N X^h(t^-, x - h) \left(\prod_{i=1}^{N} (1 - X^h(t^-, x + (i-1)h)) \right) \Bigg]$$
$$\times d(h P^h(t^-, x - h))$$
$$- \Bigg[\sum_{m=1}^{N-1} m X^h(t^-, x) \left(\prod_{i=1}^{m} (1 - X^h(t^-, x + ih)) \right)$$
$$\times X^h(t^-, x + (m+1)h)$$
$$+ N X^h(t^-, x) \left(\prod_{i=1}^{N} (1 - X^h(t^-, x + ih)) \right) \Bigg] d(h P^h(t^-, x)) \Bigg\}$$
$$+ \frac{1}{h} \Bigg\{ \Bigg[\sum_{m=1}^{N-1} m X^h(t^-, x + h) \left(\prod_{i=1}^{m} (1 - X^h(t^-, x - (i-1)h)) \right)$$
$$\times X^h(t^-, x - mh)$$

$$
\begin{aligned}
&+ NX^h(t^-, x+h) \left(\prod_{i=1}^{N} (1 - X^h(t^-, x - (i-1)h)) \right) \Bigg] \\
&\times d(hQ^h(t^-, x+h)) \\
&- \Bigg[\sum_{m=1}^{N-1} mX^h(t^-, x) \left(\prod_{i=1}^{m} (1 - X^h(t^-, x - ih)) \right) \\
&\times X^h(t^-, x - (m+1)h) \\
&+ NX^h(t^-, x) \left(\prod_{i=1}^{N} (1 - X^h(t^-, x - ih)) \right) \Bigg] d(hQ^h(t^-, x)) \Bigg\} .
\end{aligned}
$$
(8.32)

From the definition of $P^h(t, x)$, and $Q^h(t, x)$,

$$
\begin{aligned}
dX^h(t, x) = +\frac{1}{h} &\Bigg\{ \Bigg[\sum_{m=1}^{N-1} mX^h(t^-, x-h) \left(\prod_{i=1}^{m} [1 - X^h(t^-, x+(i-1)h)] \right) \\
&\times X^h(t^-, x+mh) \\
&+ NX^h(t^-, x-h) \left(\prod_{i=1}^{N} [1 - X^h(t^-, x+(i-1)h)] \right) \Bigg] \\
&\times d \left(h \sum_{k \in \mathbf{Z}} P\left(\frac{t}{h}, k \right) \mathbf{1}_{[hk, h(k+1))}(x-h) \right) \\
&- \Bigg[\sum_{m=1}^{N-1} mX^h(t^-, x) \left(\prod_{i=1}^{m} [1 - X^h(t^-, x+ih)] \right) \\
&\times X^h(t^-, x+(m+1)h) \\
&+ NX^h(t^-, x) \left(\prod_{i=1}^{N} [1 - X^h(t^-, x+ih)] \right) \Bigg] \\
&\times d \left(h \sum_{k \in \mathbf{Z}} P\left(\frac{t}{h}, k \right) \mathbf{1}_{[hk, h(k+1))}(x) \right) \Bigg\} \\
+ \frac{1}{h} &\Bigg\{ \Bigg[\sum_{m=1}^{N-1} mX^h(t^-, x+h) \left(\prod_{i=1}^{m} [1 - X^h(t^-, x-(i-1)h)] \right) \\
&\times X^h(t^-, x-mh) \\
&+ NX^h(t^-, x+h) \left(\prod_{i=1}^{N} [1 - X^h(t^-, x-(i-1)h)] \right) \Bigg] \\
&\times d \left(h \sum_{k \in \mathbf{Z}} Q\left(\frac{t}{h}, k \right) \mathbf{1}_{[hk, h(k+1))}(x+h) \right)
\end{aligned}
$$

$$- \left[\sum_{m=1}^{N-1} m X^h(t^-, x) \left(\prod_{i=1}^{m} [1 - X^h(t^-, x - ih)] \right) \right.$$
$$\times X^h(t^-, x - (m+1)h)$$
$$\left. + N X^h(t^-, x) \left(\prod_{i=1}^{N} [1 - X^h(t^-, x - ih)] \right) \right]$$
$$\times d \left(h \sum_{k \in \mathbf{Z}} Q\left(\frac{t}{h}, k \right) \mathbf{1}_{[hk, h(k+1))}(x) \right) \right\}. \tag{8.33}$$

Finally, utilizing the definition of $X^h(t, x)$ and substituting $x = hj$ gives

$$dX^h(t, x)$$
$$= + \left[\sum_{m=1}^{N-1} m X\left(\frac{t}{h}, j-1 \right) \left(\prod_{i=1}^{m} [1 - X\left(\frac{t}{h}, j+i-1 \right)] \right) X\left(\frac{t}{h}, j+m \right) \right.$$
$$\left. + N X\left(\frac{t}{h}, j-1 \right) \left(\prod_{i=1}^{N} [1 - X\left(\frac{t}{h}, j+i-1 \right)] \right) \right] d\left(P\left(\frac{t}{h}, j-1 \right) \right)$$
$$- \left[\sum_{m=1}^{N-1} m X\left(\frac{t}{h}, j \right) \left(\prod_{i=1}^{m} [1 - X\left(\frac{t}{h}, j+i \right)] \right) X\left(\frac{t}{h}, j+m+1 \right) \right.$$
$$\left. + N X\left(\frac{t}{h}, j \right) \left(\prod_{i=1}^{N} [1 - X\left(\frac{t}{h}, j+i \right)] \right) \right] d\left(P\left(\frac{t}{h}, j \right) \right)$$
$$+ \left[\sum_{m=1}^{N-1} m X\left(\frac{t}{h}, j+1 \right) \left(\prod_{i=1}^{m} [1 - X\left(\frac{t}{h}, j-(i-1) \right)] \right) X\left(\frac{t}{h}, j-m \right) \right.$$
$$\left. + N X\left(\frac{t}{h}, j+1 \right) \left(\prod_{i=1}^{N} [1 - X\left(\frac{t}{h}, j-i+1 \right)] \right) \right] d\left(Q\left(\frac{t}{h}, j+1 \right) \right)$$
$$- \left[\sum_{m=1}^{N-1} m X\left(\frac{t}{h}, j \right) \left(\prod_{i=1}^{m} [1 - X\left(\frac{t}{h}, j-i \right)] \right) X\left(\frac{t}{h}, j-m-1 \right) \right.$$
$$\left. + N X\left(\frac{t}{h}, j \right) \left(\prod_{i=1}^{N} [1 - X\left(\frac{t}{h}, j-i \right)] \right) \right] d\left(Q\left(\frac{t}{h}, j \right) \right). \tag{8.34}$$

This immediately yields the system (8.29).

As $h \to 0$, Equation (8.30) converges to the deterministic partial differential equation:

$$\frac{\partial u}{\partial t} = (1 - 2p) \frac{\partial}{\partial x} \left[\sum_{m=1}^{N-1} m u(t, x)^2 \left(1 - u(t, x) \right)^m + N u(t, x) \left(1 - u(t, x) \right)^N \right]. \tag{8.35}$$

Since

$$\sum_{m=1}^{N-1} mu^2(1-u)^m + Nu(1-u)^N$$

$$= u^2(1-u)\left[-\frac{\partial}{\partial u}\sum_{m=0}^{N-1}(1-u)^m\right] + Nu(1-u)^N$$

$$= u^2(1-u)\frac{\partial}{\partial u}\left[\frac{-1}{u}\left(1-(1-u)^N\right)\right] + Nu(1-u)^N$$

$$= u^2(1-u)\left[u^{-2}\left(1-(1-u)^N\right) - \frac{N}{u}(1-u)^{N-1}\right] + Nu(1-u)^N$$

$$= (1-u)\left(1-(1-u)^N\right) - Nu(1-u)^N + Nu(1-u)^N$$

$$= (1-u) - (1-u)^{N+1}. \tag{8.36}$$

The above arguments lead to the following result.

Theorem 8.3. *Let $p \neq 1/2$. As $h \to 0$, the solution $X^h(t,x)\,dx$ of (8.30) converges weakly to $u(t,x)\,dx$, where $u(t,x)$ is a weak solution of the nonlinear Cauchy problem*

$$\frac{\partial u}{\partial t} + (2p-1)\frac{\partial G(u)}{\partial x} = 0. \tag{8.37}$$

$$u(0,x) = u_0(x) = b\mathbf{1}_{(-\infty,0]}(x) + a\mathbf{1}_{(0,\infty)}(x), \tag{8.38}$$

with some $0 \leq a < b < \infty$ and

$$G(u) = \left[\left(1-u(t,x)\right) - \left(1-u(t,x)\right)^{N+1}\right]. \tag{8.39}$$

Moreover, for all t,x, we have $a \leq u(t,x) \leq b$.

8.5 SHOCK AND RAREFACTION WAVE SOLUTIONS FOR THE RIEMANN PROBLEM FOR CONSERVATION LAWS

The nonlinear hyperbolic equations (8.25) and (8.37) describing the density profiles for the queuing networks in Sections 8.3 and 8.4 are special cases of general conservation laws (Smoller, 1994) of the form

$$\frac{\partial u}{\partial t} + \frac{\partial H(u)}{\partial x} = 0. \tag{8.40}$$

and in the case of initial conditions of the form

$$u(0,x) = u_0(x) = u_l\mathbf{1}_{(-\infty,0]}(x) + u_r\mathbf{1}_{(0,\infty]}(x), \tag{8.41}$$

where u_l and u_r are constants (so called Riemann problem), they can be solved explicitly under some extra conditions on function H.

Let us recall (Smoller, 1994) that a bounded and measurable function $u(t, x)$ is called a (weak) solution of the initial-value problem

$$\frac{\partial u}{\partial t} + \frac{\partial H(u)}{\partial x} = 0, \qquad u(0, x) = u_0(x), \tag{8.42}$$

with bounded and measurable initial data u_0 if

$$\int_{t \geq 0} \int_{\mathbf{R}} \left(u\phi_t + H(u)\phi_x \right) dx \, dt + \int_{t=0} u_0 \phi \, dx = 0 \tag{8.43}$$

In general, solutions are not unique unless additional assumptions, such as the entropy condition mentioned below, are satisfied.

The solutions of the Riemann problem (8.40–8.41) are invariant under hyperbolic rescaling, that is, for every constant $\lambda > 0$

$$u_\lambda(t, x) = u(\lambda t, \lambda x)$$

is a solution whenever u is. Thus, one looks for the solutions of the form

$$u(t, x) = v(x/t) \tag{8.44}$$

This gives rise to three types of local behavior of the solutions of u:

- $u(t, x)$ is constant;

- $u(t, x)$ is a *shock wave* of the form

$$u(t, x) = u_0 \mathbf{1}_{(-\infty, Vt)}(x) + u_1 \mathbf{1}_{[Vt, \infty)}(x), \tag{8.45}$$

  traveling with the velocity

$$V = \frac{H(u_0) - H(u_1)}{u_0 - u_1}.$$

  For the sake of uniqueness, one adds here the entropy condition $H'(u_0) > V > H'(u_1)$.

- $u(t, x)$ is a continuous *rarefaction wave* of the form (8.44) where v satisfies the ordinary differential equation

$$v'(\xi)\big(H'(v(\xi)) - \xi\big) = 0. \tag{8.46}$$

We will apply the above standard observations to the case of the GS and multiple servers queuing networks starting with the latter because it is somewhat simpler (see also Margolius, 1999 for other approaches to multiserver queues). The animated graphics files depicting the time evolution of density profiles for both types of networks can be seen at our website: `http://www.academic.csuohio.edu/bmargolius/waves/wavesa.htm`.

Density profiles for the multiple servers network. The network was described in Section 4, and we will study it here in the special case of totally asymmetric $p = 1$ and the initial condition (8.41) where $u_l = 1$ and $u_r = 0$; other situations with Riemann-type data can treated in a similar fashion and will be analyzed elsewhere. Therefore, from now on, in this subsection

$$H(u) = \left[(1 - u) - (1 - u)^{N+1} \right]. \tag{8.47}$$

Thus, for ξ such that $v'(\xi) \neq 0$, Equation (8.46) can be written in the form

$$-1 + (N + 1)(1 - v)^N = \xi, \tag{8.48}$$

so that the solution of the Riemann problem is

$$u(t, x) = \begin{cases} 1, & for\ x < -t; \\ 1 - [(x/t + 1)/(N + 1)]^{1/N}, & for\ -t \leq x < Nt; \\ 0, & for\ Nt \leq x. \end{cases} \tag{8.49}$$

Density profiles for the GS network. This case is slightly more difficult as uniqueness questions arise because of the bifurcations. For the GS network, we examine the solution with initial condition $u_l = 2/(m + 1)$, $u_r = 0$, that is, we begin with an average of $m - 1$ customers for every two servers to the left of zero and no servers to the right of zero. Proceeding as in the multiserver case, we have

$$H(u) = u(1 - u)^m, \tag{8.50}$$

so for ξ such that $v(\xi) \neq 0$, Equation (8.46) can be written in the form

$$(1 - u)^{m-1}(1 - (m + 1)u) = \xi. \tag{8.51}$$

Therefore, for the GS network, the solution of the Riemann problem bifurcates at (x_b, u_b), where

$$x_b = -t \left(\frac{m - 1}{m + 1} \right)^{m-1},$$

$$u_b = 2/(m + 1). \tag{8.52}$$

The figure below illustrates the case where $m = 4$, but the solutions are similar to those shown for $m > 1$. When $m = 1$, we have a simple asymmetric exclusion process.

The lower branch is consistent with the entropy condition (see, e.g., Smoller (1994), p.). Define $g(u) = t(1 - u)^{m-1}(1 - (m + 1)u)$ for $0 < u \leq 2/(m + 1)$. Hence, in the case of a GS network, the solution of the Riemann problem satisfying the entropy condition is

$$u(t, x) = \begin{cases} 2/(m + 1) & for\ x < -t \left(\frac{m-1}{m+1} \right)^{m-1}; \\ g^{-1}(x) & for\ -t \left(\frac{m-1}{m+1} \right)^{m-1} \leq x < t; \\ 0 & for\ t \leq x. \end{cases} \tag{8.53}$$

Note that at location $x = 0$, for $t > 0$, there is an average of m customers for every one server, i.e., $u(t, 0) = 1/(m + 1), \forall t > 0$; for $x < 0$ the average server does not have enough customers waiting to begin service, and for $x > 0$, the average server has a queue longer than the m customers required to begin service. We will analyze the behavior of the GS network more thoroughly elsewhere.

CHAPTER 9

Nonlinear Diffusion Equations

9.1 HYPERBOLIC EQUATIONS

For initial conditions not of Riemann type, in particular those with integrable data, or for more general random initial conditions, obtaining solutions of the conservations law is not a simple matter, even in approximate fashion. The usual approach then is to consider a parabolic regularization (the viscosity method) by considering the *nonlinear diffusion* equations

$$\frac{\partial u}{\partial t} + \frac{\partial H(u)}{\partial x} = \epsilon \mathcal{L} u, \qquad u(0,x) = u_0(x), \tag{9.1}$$

where $\mathcal{L}$ is a dissipative operator of elliptic type, like e.g. the Laplacian. Then, of course, with the exception of the quadratic case giving rise to the Burgers equation, one cannot count on finding explicit solutions, but two types of asymptotic results can be used as approximations.

The first kind provides the large time asymptotics of the regularized conservation laws and the second kind gives a Monte Carlo method of solving them through the interacting diffusions scheme (so-called propagation of chaos). We will briefly describe the two approaches.

Asymptotics for nonlinear diffusion equations. Not surprisingly, given the decay of their solution in time, the large time asymptotic behavior for parabolically regularized conservation laws is dictated by the asymptotic behavior of the nonlinearity $H(u)$ at points where the function is small. Therefore, we have the following asymptotic results for regularized versions of the hyperbolic equations:

Theorem 9.1. *Let $\epsilon > 0, m \geq 1$ and $u(t,x)$ be a positive weak solution of the Cauchy problem*

$$\frac{\partial u}{\partial t} + (2p-1)\frac{\partial F(u)}{\partial x} = \epsilon \frac{\partial^2 u}{\partial x^2}, \qquad 1 \geq u(0,x) = u_0(x) \geq 0, \tag{9.2}$$

with $F(u) = [u(1-u)^m]$. Then

DOI: 10.1201/9781003216759-9

(i) If $u_0 \in L^1(\mathbf{R})$ then u has the same large time asymptotics as the solution of the linear diffusion equation

$$\frac{\partial u}{\partial t} + (2p - 1)\frac{\partial u}{\partial x} = \epsilon \frac{\partial^2 u}{\partial x^2}, \qquad 1 \geq u(0, x) = u_0(x) \geq 0, \tag{9.3}$$

or more precisely

$$\|u(t, x) - U(t, x)\|_1 \to 0 \quad \text{as} \quad t \to \infty, \tag{9.4}$$

where $U(t, x) = (g * u_0)(t, x - (2p - 1)t)$ and $g(t, x) = (4\pi t)^{-1/2} \exp(-|x|^2/(4t))$ is the standard Gaussian kernel.

(ii) If $1 - u_0 \in L^1(\mathbf{R})$ then:

In the case $m = 1$, u has the same large time asymptotics as the solution of the linear diffusion equation.

In the case $m = 2$, u has the same large time asymptotics as the self-similar source solution of the Burgers equations or more precisely, for each $p > 1$

$$t^{(1-1/p)/2}\|u(t, x) - U_M(t, x)\|_p \to 0 \quad \text{as} \quad t \to \infty, \tag{9.5}$$

where

$$U_M(t, x) = t^{-1/2} \exp(-x^2/(4t))\left(K(M) + \frac{1}{2}\int_0^{x/(2\sqrt{t})} \exp(-\xi^2/4)\, d\xi\right)^{-1}, \tag{9.6}$$

and $U_M(t, x) \to M\delta(x)$ as $t \to 0$ with $M = \|u_0\|_1$.

In the case $m \geq 3$, u has the same large time asymptotics as the solution of the heat equations or more precisely, for each $p > 1$ there exists a constant C such that

$$\|u(t, x) - U(t, x)\|_p \leq Ct^{-(1-1/p)/2}, \tag{9.7}$$

where $U(t, x) = (g * u_0)(t, x)$.

Proof. **(Sketch of the Proof).** By the results of Escobedo and Zuazua (1991), Escobedo, Velazquez and Zuazua (1993) (see also Biler, Karch and Woyczynski, 1999 for other regularizations of conservation laws) the asymptotic behavior of the solutions of the conservation laws depends on the asymptotic behavior of the nonlinearity H at its small values. So, for $H(u) = (2p-1)F(u) = (2p-1)u(1-u)^m$,

$$\lim_{u \to 0} \frac{(2p-1)F(u)}{u} = 2p - 1, \tag{9.8}$$

and

$$\lim_{u \to 1} \frac{(2p-1)F(u)}{(1-u)^m} = 2p - 1. \tag{9.9}$$

The first condition, together with the standard step removing the drift term in the linear diffusion equation gives (i) and the case $m = 1$ in the second condition gives the first part of (ii).

The critical case $m = 2$ yields the Burgers equation type asymptotics claimed in the second part of (ii), and the supercritical case $m \geq 3$ where the effect of the nonlinear convection term disappears in the limit. ☐

Theorem 9.2. *Let $\epsilon > 0, N \geq 1$ and $u(t,x)$ be a positive weak solution of the Cauchy problem*

$$\frac{\partial u}{\partial t} + (2p-1)\frac{\partial G(u)}{\partial x} = \epsilon \frac{\partial^2 u}{\partial x^2}, \qquad 1 \geq u(0,x) = u_0(x) \geq 0, \qquad (9.10)$$

with $G(u) = (1-u) - (1-u)^{N+1}$. Then if either $u_0 \in L^1(\mathbf{R})$ or $1 - u_0 \in L^1(\mathbf{R})$ then u has the same large time asymptotics as the solution of the linear diffusion equation

$$\frac{\partial u}{\partial t} + (2p-1)\frac{\partial u}{\partial x} = \epsilon \frac{\partial^2 u}{\partial x^2}, \qquad 1 \geq u(0,x) = u_0(x) \geq 0, \qquad (9.11)$$

or more precisely

$$\|u(t,x) - U(t,x)\|_1 \to 0 \quad \text{as} \quad t \to \infty, \qquad (9.12)$$

where $U(t,x) = (g * u_0)(t, x - (2p-1)t)$ and $g(t,x) = (4\pi t)^{-1/2} \exp(-|x|^2/(4t))$ is the standard Gaussian kernel.

Proof. **(Sketch of the Proof).** The proof of this result relies on the same asymptotic results that were employed in the proof of Theorem 9.1. However, in this case $H(u) = (2p-1)[(1-u) - (1-u)^{N+1}]$ which has the linear asymptotics at both $u = 0$ and $u = 1$. Therefore, the result follows by the usual reduction to the heat equation. □

9.2 NONLINEAR DIFFUSION APPROXIMATIONS

For initial conditions not of Riemann type, in particular those with integrable data, or for more general random initial conditions, obtaining solutions of the conservations law is not a simple matter, even in approximate fashion. The usual approach then is to consider a parabolic regularization (the viscosity method) by considering the *nonlinear diffusion* equations

$$\frac{\partial u}{\partial t} + \frac{\partial H(u)}{\partial x} = \epsilon \mathcal{L}u, \qquad u(0,x) = u_0(x), \qquad (9.13)$$

where $\mathcal{L}$ is a dissipative operator of elliptic type, like e.g. the Laplacian. Then, of course, with the exception of the quadratic case giving rise to the Burgers equation, one cannot count on finding explicit solutions, but two types of asymptotic results can be used as approximations.

The first kind provides the large time asymptotics of the regularized conservation laws and the second kind gives a Monte Carlo method of solving them via the interacting diffusions scheme (so-called propagation of chaos). We will briefly describe the two approaches.

Asymptotics for nonlinear diffusion equations. Not surprisingly, given the decay of their solution in time, the large time asymptotic behavior for parabolically regularized conservation laws is dictated by the asymptotic behavior of the nonlinearity $H(u)$ at points where the function is small. Therefore, we have the following asymptotic results for regularized versions of (9.25) and (9.37):

Theorem 9.3. *Let $\epsilon > 0, m \geq 1$ and $u(t,x)$ be a positive weak solution of the Cauchy problem*

$$\frac{\partial u}{\partial t} + (2p-1)\frac{\partial F(u)}{\partial x} = \epsilon \frac{\partial^2 u}{\partial x^2}, \qquad 1 \geq u(0,x) = u_0(x) \geq 0, \qquad (9.14)$$

with $F(u) = [u(1-u)^m]$. Then

(i) If $u_0 \in L^1(\mathbf{R})$, then u has the same large time asymptotics as the solution of the linear diffusion equation

$$\frac{\partial u}{\partial t} + (2p-1)\frac{\partial u}{\partial x} = \epsilon \frac{\partial^2 u}{\partial x^2}, \qquad 1 \geq u(0,x) = u_0(x) \geq 0, \qquad (9.15)$$

or more precisely

$$\|u(t,x) - U(t,x)\|_1 \to 0 \quad \text{as} \quad t \to \infty, \qquad (9.16)$$

where $U(t,x) = (g * u_0)(t, x - (2p-1)t)$ and $g(t,x) = (4\pi t)^{-1/2} \exp(-|x|^2/(4t))$ is the standard Gaussian kernel.

(ii) If $1 - u_0 \in L^1(\mathbf{R})$ then:

In the case $m = 1$, u has the same large time asymptotics (9.56–9.57) as the solution of the linear diffusion equation.

In the case $m = 2$, u has the same large time asymptotics as the self-similar source solution of the Burgers equations or more precisely, for each $p > 1$

$$t^{(1-1/p)/2}\|u(t,x) - U_M(t,x)\|_p \to 0 \quad \text{as} \quad t \to \infty, \qquad (9.17)$$

where

$$U_M(t,x) = t^{-1/2} \exp(-x^2/(4t))\left(K(M) + \frac{1}{2}\int_0^{x/(2\sqrt{t})} \exp(-\xi^2/4)\, d\xi\right)^{-1}, \quad (9.18)$$

and $U_M(t,x) \to M\delta(x)$ as $t \to 0$ with $M = \|u_0\|_1$.

In the case $m \geq 3$, u has the same large time asymptotics as the solution of the heat equations or more precisely, for each $p > 1$, there exists a constant C such that

$$\|u(t,x) - U(t,x)\|_p \leq Ct^{-(1-1/p)/2}, \qquad (9.19)$$

where $U(t,x) = (g * u_0)(t,x)$.

Proof. **(Sketch of the Proof).** By the results of Escobedo and Zuazua (1991), Escobedo, Velazquez and Zuazua (1993) (see also Biler, Karch and Woyczynski (1999) for other regularizations of conservation laws) the asymptotic behavior of the solutions of the conservation laws (54) depends on the asymptotic behavior of the nonlinearity H at its small values. Therefore, for $H(u) = (2p-1)F(u) = (2p-1)u(1-u)^m$,

$$\lim_{u \to 0} \frac{(2p-1)F(u)}{u} = 2p-1, \qquad (9.20)$$

and

$$\lim_{u \to 1} \frac{(2p-1)F(u)}{(1-u)^m} = 2p-1. \qquad (9.21)$$

The first condition (61), together with the standard step removing the drift term in the linear diffusion equation gives (i), and the case $m = 1$ in the second condition (62) gives the first part of (ii).

The critical case $m = 2$ in (62) yields the Burgers equation-type asymptotics claimed in the second part of (ii), and the supercritical case $m \geq 3$ where the effect of the nonlinear convection term disappears in the limit. □

Theorem 9.4. *Let $\epsilon > 0, N \geq 1$ and $u(t,x)$ be a positive weak solution of the Cauchy problem*

$$\frac{\partial u}{\partial t} + (2p-1)\frac{\partial G(u)}{\partial x} = \epsilon\frac{\partial^2 u}{\partial x^2}, \qquad 1 \geq u(0,x) = u_0(x) \geq 0, \qquad (9.22)$$

with $G(u) = (1-u) - (1-u)^{N+1}$. Then if either $u_0 \in L^1(\mathbf{R})$ or $1 - u_0 \in L^1(\mathbf{R})$, then u has the same large time asymptotics as the solution of the linear diffusion equation

$$\frac{\partial u}{\partial t} + (2p-1)\frac{\partial u}{\partial x} = \epsilon\frac{\partial^2 u}{\partial x^2}, \qquad 1 \geq u(0,x) = u_0(x) \geq 0, \qquad (9.23)$$

or more precisely

$$\|u(t,x) - U(t,x)\|_1 \to 0 \quad \text{as} \quad t \to \infty, \qquad (9.24)$$

where $U(t,x) = (g * u_0)(t, x - (2p-1)t)$ and $g(t,x) = (4\pi t)^{-1/2}\exp(-|x|^2/(4t))$ is the standard Gaussian kernel.

Proof. **(Sketch of the Proof).** The proof of this result relies on the same asymptotic results that were employed in the proof of Theorem 6.1. However, in this case $H(u) = (2p-1)[(1-u) - (1-u)^{N+1}]$ which has the linear asymptotics at both $u = 0$ and $u = 1$. Therefore, the result follows by the usual reduction to the heat equation. □

Interacting diffusions approximations for nonlinear diffusion equations. This section discusses a possibility of a Monte Carlo-type approximation for solutions of nonlinear diffusion equations of the type that arise as parabolic regularizations (54) of conservation laws of the encountered in Theorems 3.1, 4.1, and 4.2. The idea is to use the following scheme known as the propagation of chaos result and depends on the construction of the so-called nonlinear McKean process for our equations.

The basic observation is that if the regularizing operator $\mathcal{L}$ is the infinitesimal generator of a Lévy process then the Equation (54) (say, $\epsilon = 1$) can be formally interpreted as a "Fokker–Planck–Kolmogorov equation" for a "nonlinear" diffusion process in the McKean's sense. Indeed, consider a Markov process $X(t)$, $t \geq 0$, which is a solution of the stochastic differential equation

$$dX(t) = dS(t) - u^{-1}H(u(X(t), t))\, dt, \qquad (9.25)$$
$$X(0) \sim u_0(x)\, dx \ \text{ in law},$$

where $S(t)$ is the Lévy process with generator $-\mathcal{L}$. Assuming that $X(t)$ is a unique solution of (9.28), we see that the measure-valued function $v(dx, t) = P(X(t) \in dx)$ satisfies the weak forward equation

$$\frac{d}{dt}\langle v(t), \eta \rangle = \langle v(t), \widetilde{\mathcal{L}}_{u(t)}\eta \rangle, \ \eta \in \mathcal{S}(\boldsymbol{b}R^n), \qquad (9.26)$$
$$v(0) = u(x, 0)\, dx$$

with $\widetilde{\mathcal{L}}_u = -\mathcal{L} + u^{-1}H(u) \cdot \nabla$. On the other hand $u(dx, t) = u(x, t)\, dx$ also solves (9.29) since

$$\frac{d}{dt}\langle u(t), \eta \rangle = \langle -\mathcal{L}u - \nabla \cdot H(u), \eta \rangle = \langle u, (-\mathcal{L} + u^{-1}H(u) \cdot \nabla)\eta \rangle$$

so that $v(dx, t) = u(dx, t)$ and, by uniqueness, u is the density of the solution of (66).

The above construction makes a possible approximation of solutions of equations of the form (54) through finite systems of interacting diffusions. To illustrate our point, we will formulate this Monte Carlo algorithm in the special and well-known Burgers equation case where $\mathcal{L} = \Delta$, is the usual Laplacian and the nonlinearity $H(x) = x^2$ is quadratic. The more general results needed for the analysis of GS and multiserver queuing networks are under development (Calderoni and Pulvirenti, 1983; Sznitman, 1991; Zhang, 1995; Funaki and Woyczynski, 1998; Woyczynski, 1998; Biler, Funaki and Woyczynski, 2000; Margolius, Subramanian and Woyczynski, 2000, for more details on the subject).

For each $n \in \mathbf{N}$, let us introduce independent, symmetric, real-valued standard Brownian motion processes $\{S^i(t), \ i = 1, 2, \ldots, n\}$, and let $\delta_\epsilon(x) := (2\pi\epsilon)^{-1/2}\exp\left[-x^2/2\epsilon\right], \epsilon > 0$, be a regularizing kernel. Consider a system of n interacting particles with positions $\{X^i(t)\}_{i=1,\ldots,n} \equiv \{X^{i,n,\epsilon}(t)\}_{i=1,\ldots,n}$, and the corresponding measure-valued process (empirical distribution) $\bar{X}^n(t) \equiv \bar{X}^{n,\epsilon}(t) := \frac{1}{n}\sum_{i=1}^n \delta(X^{i,n,\epsilon}(t))$, with the dynamics provided by the system of regularized singular stochastic differential equations

$$dX^i(t) = dS^i(t) + \frac{1}{n}\sum_{j \neq i}\delta_\epsilon(X^i(t) - X^j(t))\, dt, \quad i = 1, \ldots, n, \qquad (9.27)$$

and the initial conditions $X^i(0) \sim u_0(x)$ (in distribution, thus, $u_0 \in L_1$ here). Then, for each $\epsilon > 0$, the empirical process $\bar{X}^{n,\epsilon}(t) \Longrightarrow u^\epsilon(x,t)\,dx$, in probability, as $n \to \infty$, where $\Rightarrow$ denotes the weak convergence of measures, and the limit density $u^\epsilon \equiv u^\epsilon(x,t)$, $t > 0$, $x \in \mathbf{R}$, satisfies the regularized Burgers equation $u_t^\epsilon + \left(\frac{1}{2}(\delta_\epsilon * u^\epsilon) \cdot u^\epsilon\right)_x = \Delta u^\epsilon$ with the initial condition $u(0,x) = u_0(x)$. The speed of convergence is controlled (Bossy and Talay, 1996). Moreover, under some additional technical conditions, for a class of test functions ϕ, $E|\langle \bar{X}^{n,\epsilon(n)}(t) - u(t), \phi\rangle| \longrightarrow 0$, as $n \to \infty, \epsilon(n) \to 0$, where $u(t) = u(x,t)$ is a solution of the nonregularized Burgers equation $u_t + (u^2)_x = \Delta u$ with the initial condition $u(0,x) = u_0(x)$.

9.3   NONLINEAR PROCESSES

Interacting diffusions approximations for nonlinear diffusion equations. This section discusses a possibility of a Monte Carlo-type approximation for solutions of nonlinear diffusion equations of the type that arise as parabolic regularizations of conservation laws that encountered before. The idea is to use the following scheme known as the propagation of chaos result and depends on the construction of the so-called nonlinear McKean process for our equations.

The basic observation is that if the regularizing operator $\mathcal{L}$ is the infinitesimal generator of a Lévy process, then the parabolic equation of the previous chapter (say, $\epsilon = 1$) can be formally interpreted as a "Fokker–Planck–Kolmogorov equation" for a "nonlinear" diffusion process in the McKean's sense. Indeed, consider a Markov process $X(t)$, $t \geq 0$, which is a solution of the stochastic differential equation

$$dX(t) = dS(t) - u^{-1}H(u(X(t),t))\,dt, \tag{9.28}$$
$$X(0) \sim u_0(x)\,dx \quad \text{in law},$$

where $S(t)$ is the Lévy process with generator $-\mathcal{L}$. Assuming that $X(t)$ is a unique solution of (9.28), we see that the measure-valued function $v(dx,t) = P(X(t) \in dx)$ satisfies the weak forward equation

$$\frac{d}{dt}\langle v(t),\eta\rangle = \langle v(t),\widetilde{\mathcal{L}}_{u(t)}\eta\rangle, \quad \eta \in \mathcal{S}(\mathbf{R}^n), \tag{9.29}$$
$$v(0) = u(x,0)\,dx$$

with $\widetilde{\mathcal{L}}_u = -\mathcal{L} + u^{-1}H(u) \cdot \nabla$. On the other hand $u(dx,t) = u(x,t)\,dx$ also solves (9.29) since

$$\frac{d}{dt}\langle u(t),\eta\rangle = \langle -\mathcal{L}u - \nabla \cdot H(u),\eta\rangle = \langle u,(-\mathcal{L} + u^{-1}H(u) \cdot \nabla)\eta\rangle$$

so that $v(dx,t) = u(dx,t)$ and, by uniqueness, u is the density of the solution of (1).

9.4   INTERACTING DIFFUSIONS AND MONTE-CARLO METHODS

The above construction makes a possible approximation of solutions of the parabolic equations through finite systems of interacting diffusions. To illustrate our point, we will formulate this Monte Carlo algorithm in the special, and well-known Burgers equation case where $\mathcal{L} = \Delta$, is the usual Laplacian and the nonlinearity $H(x) = x^2$ is quadratic. The more general results needed for the analysis of GS and multi-server queuing networks are under development (Calderoni and Pulvirenti, 1983; Sznitman, 1991; Zhang, 1995; Funaki and Woyczynski, 1998; Woyczynski, 1998; Biler, Funaki and Woyczynski, 2000; Margolius, Subramanian and Woyczynski, 2000, for more details on the subject).

For each $n \in \mathbf{N}$, let us introduce independent, symmetric, real-valued standard Brownian motion processes $\{S^i(t), i = 1, 2, \ldots, n\}$, and let $\delta_\epsilon(x) := (2\pi\epsilon)^{-1/2} \exp\left[-x^2/2\epsilon\right], \epsilon > 0$, be a regularizing kernel. Consider a system of n interacting particles with positions $\{X^i(t)\}_{i=1,\ldots,n} \equiv \{X^{i,n,\epsilon}(t)\}_{i=1,\ldots,n}$, and the corresponding measure-valued process (empirical distribution) $\bar{X}^n(t) \equiv \bar{X}^{n,\epsilon}(t) := \frac{1}{n}\sum_{i=1}^n \delta(X^{i,n,\epsilon}(t))$, with the dynamics provided by the system of regularized singular stochastic differential equations

$$dX^i(t) = dS^i(t) + \frac{1}{n}\sum_{j \neq i} \delta_\epsilon(X^i(t) - X^j(t))\, dt, \quad i = 1, \ldots, n, \qquad (9.30)$$

and the initial conditions $X^i(0) \sim u_0(x)$ (in distribution, thus, $u_0 \in L_1$ here). Then, for each $\epsilon > 0$, the empirical process $\bar{X}^{n,\epsilon}(t) \Longrightarrow u^\epsilon(x,t)\, dx$, in probability, as $n \to \infty$, where $\Rightarrow$ denotes the weak convergence of measures, and the limit density $u^\epsilon \equiv u^\epsilon(x,t)$, $t > 0$, $x \in \mathbf{R}$, satisfies the regularized Burgers equation $u^\epsilon_t + (\frac{1}{2}(\delta_\epsilon * u^\epsilon) \cdot u^\epsilon)_x = \Delta u^\epsilon$. with the initial condition $u(0,x) = u_0(x)$. The speed of convergence is controlled (Bossy and Talay 1996). Moreover, under some additional technical conditions, for a class of test functions ϕ, $E|\langle \bar{X}^{n,\epsilon(n)}(t) - u(t), \phi\rangle| \longrightarrow 0$, as $n \to \infty, \epsilon(n) \to 0$, where $u(t) = u(x,t)$ is a solution of the nonregularized Burgers equation $u_t + (u^2)_x = \Delta u$ with the initial condition $u(0,x) = u_0(x)$.

The Remarkable Bernoulli Family[1]

BERNOULLI. Originally from Antwerp, the family became citizens of Basel in 1622. Coat of arms: In silver, a tri-partite green branch, each having seven (sometimes nine) leaves. The progenitor was Jakob (d.1583), who in 1570 had fled from Antwerp to Frankfurt/Main, thus escaping Count Alba's persecution of heretics. Jakob Bernoulli, a wholesale grocer and tradesman, had 17 children. One line of his descendants stayed in Frankfurt, where it is still flourishing; others settled in Hamburg, Cologne, Breslau (Wrocław), and Basel.

1. **Jakob** (1598–1634) Grandson of progenitor Jakob, druggist and grocer, was made a citizen of Basel in 1622.

2. **Niklaus** (1623–1708) Son of 1. Elected member of the board of the Saffron's Guild; representing the guild on the city's Great Council [legislative] in 1668.

3. **Niklaus** (1662–1716) Son of 2, painter, elected to the Small Council [executive] in 1705.

The family's reputation for its outstanding impact on science, especially mathematics, was established first by:

4. **Jakob**[2] (1654–1705) Brother of 3. Professor of Mathematics at Basel University. After graduating in theology in 1676, he traveled widely across Switzerland, France, the Netherlands, and England, where he made contact with the most prominent mathematicians. On his return to Basel in 1682, he inaugurated lectures on experimental physics and, in 1687, was appointed to the Chair of Mathematics. Inspired by, yet independently from, Leibniz, he explored the infinitesimal calculus. He published studies on the logarithmic spiral, the loxodromes, infinite fractions, infinite series (on which occasion the

---

[1] Adapted, with permission, from THE BERNOULLI NEWS, June 1994, 15–17.
[2] Author of *Ars Conjectandi*

so-called Bernoulli numbers were discovered), etc., as well as studies on the isoperimetric problem—hereby laying the foundation for variational calculus. In 1701, he became a member of the Berlin Academy.

5. **Johann** (1667–1748) Younger brother of 4. Also, Professor of Mathematics and even more renowned for his contributions. He had studied medicine as well, qualifying in this discipline at Basel University in 1694. The next year he became Professor of Mathematics and Physics at Groningen. In 1705, he succeeded his brother to the chair at Basel University. Listing all his achievements in mathematics would require a comprehensive survey of the whole higher analysis; suffice it to name his evaluation, using differential calculus, of the limits of quotients whose numerator and denominator both tend to zero, invention of calculus for exponential functions, most integration methods (together with Leibniz), etc. He was elected a corresponding member of the Academies of Paris (1699), Berlin (1701), London (1712), Bologna (1724), and Petersburg (1725). Among his students were his own sons Niklaus, Daniel, and Johann—and also Leonhard Euler.

6. **Niklaus**[3] (1687–1759) Son of 3. Mathematician, lawyer, and philosopher. His favorite field of research was the theory of infinite series. From 1716 to 1719, he was Professor of Mathematics at Padua University; in 1722, Basel University appointed him as Professor of Logic, and as Professor of Codified and Feudal Law in 1731. Member of the Academies of Berlin, London, and Bologna.

7. **Niklaus II** (1695–1726) Son of 5. Mathematician and lawyer. Professor of Law at Bern University in 1723. In 1725, he left (with his brother Daniel) for Petersburg's newly founded Academy, where he suffered an early death on July 26, 1726. His contributions to mathematics lie above all in the field of integral and differential calculus.

8. **Daniel** (1700–1782) Son of 5. Mathematician, physicist, medical doctor, and botanist, studied mathematics with his father and his brother Niklaus from the age of 11, and later studied medicine in Basel, Heidelberg, and Venice. In 1725, he left for Petersburg with his brother Niklaus. Between 1725 and 1757, he was ten times awarded prizes for his mathematical work by the Paris Academy—on one occasion jointly with his father (1734); on another with his youngest brother, Johann. In 1733, he was made Professor of Anatomy and Botany at Basel University. His scientific achievements reach out into the most diverse areas of mathematics, physics, astronomy, etc. Member of the Academies of Petersburg, Berlin, London, and Paris.

9. **Johann II** (1710–1790) Son of Johann (5.). Mathematician and lawyer. After obtaining his doctorate in law, 1732, he joined his older brother Daniel in Petersburg, but returned with him to Basle the following year. In 1743, he

---

[3] Editor of *Ars Conjectandi*. He also applied the Law of Large Numbers to longevity data.

was given the Chair of Rhetoric at Basel University, which he exchanged for a Chair of Mathematics in 1748. Four of his studies were awarded prizes by the Paris Academy.

10. **Johann III** (1744–1807) Eldest son of 9. Astronomer and mathematician. Appointed by the Berlin Academy in 1763, he became Director of the Berlin Observatory in 1767. Member of the Academies of Paris, Petersburg, Rome, London, and Stockholm. Director of the Berlin Academy.

11. **Daniel** (1751–1834) Second son of 9. Doctor of Medicine, Professor of Rhetoric, followed by professorships in physics and medicine; finally, also Major-Domo of the Dean of the Basel Cathedral.

12. **Jakob II** (1759–1789) Youngest son of Johann II (9.). Mathematician and physicist, he served as secretary to the Imperial Ambassador in Venice, from 1779. In 1786, he joined the Petersburg Academy. His research touched on problems of theoretical mechanics.

The descendants of the fourth son of Johann II, Emanuel, live in Venice and Petersburg. From among the descendants of the fifth son, the pharmacist Niklaus, mention should be made of:

13. **Leonhard** (1791–1871) Councillor of the city of Basel.

14. **Niklaus** (1793–1876) President of the Criminal Court.

15. **August** (1839–1921) Son of 13. Ph.D., renowned historian.

16. **August** (1879–) Son of 15. Professor of Physical Chemistry at Basel University.

17. **Hans** (1876–) Grandson of 14. Architect and Professor at ETH Zürich.

Among the descendants of Daniel (11), quite a few have been important figures:

18. **Christoph** (1782–1863) Biologist and technologist; founder and director of the Philotechnic Institute; Professor of Natural History at Basel University.

19. **Carl Christoph** (1861–) Grandson of 18. Ph.D., chief librarian of the Basel Great Council.

20. **Carl Albrecht** (1863–) Licentiate in Theology, novelist and dramatist. His works are listed in the *Schweizerisches Zeitgenossen Lexikon (Swiss Lexicon of Contemporaries)*.

21. **Johannes** (1864–1920) Ph.D., Director of the Swiss National Library in Bern, 1895–1908.

22. **Eduard** (1867–) Brother of 21. Professor of History of Music, Zürich University.

23. **Hieronymus** (1745–1829) Great-nephew of 4 and 5. Biologist.

24. **Karl Gustav** (1834–1878) Biologist.

25. **Johann Jakob** (1831–1920) Ph.D., Professor of Archaeology at Basel University.

The Bernoulli family is still thriving; Daniel Bernoulli is currently Professor of Geology at ETH Zürich.

Bibligraphical Note: The above biographical data have been taken, with the help from M.G. SOLAND of ETH Zürich in preparation of the English version, from

- [1] H.TÜRLER ET AL., Eds., *Historisch-biographisches Lexikon der Schweiz*, 7 vols, Neuchâtel, 1921 - 1934, Vol. 2 (1924).

All Bernoullis active and successful in mathematics and/or any other area of knowledge were given an entry, chronologically listed up to 1922. For detailed information on specific achievements of this famous Swiss family consult

- [2] C. C. GILLESPIE, Ed., *Dictionary of Scientific Biography*, Vol. 2, Charles Scribner and Sons, New York 1970.

Bibliography

[Benassi (1984)] Benassi A., Fouque J.P., Hydrodynamical limit for the asymmetric simple exclusion process, *Ann. Probab.* **15** (1984), 546–560.

[Biler (2000)] Biler P., Funaki T., Woyczynski W.A., Interacting particle approximation for nonlocal quadratic evolution problems, *Prob. Math. Stat.* **20** (2000), 1–23.

[Biler (1999)] Biler P., Karch G., Woyczynski W.A., Asymptotics of multifractal conservation laws, *Studia Math.* **135** (1999), 231–252.

[Biler (2000)] Biler P., Karch G., Woyczynski W.A., Multifractal and Lévy conservation laws, *Comptes Rendus Acad. Sci. (Paris)* (2000), 1–4.

[Bossy (1996)] Bossy M., Talay D., Convergence rate for the approximation of the limit law of weakly interacting particles: Application to the Burgers equation, *Ann. Appl. Probab.* **6** (1996), 818–861.

[Calderoni (1983)] Calderoni P., Pulvirenti M., Propagation of chaos for Burgers' equation, *Ann. Inst. H. Poincaré - Phys. Th.* **39** (1983), 85–97.

[Escobedo (1991)] Escobedo M., Zuazua E., Large time behavior for convection-diffusion equations in bR^N, *J. Funct. Anal.* **100** (1991), 119–161.

[Escobedo (1993)] Escobedo M., Vázquez J.L., Zuazua E., Asymptotic behaviour and source-type solutions for a diffusion-convection equation, *Arch. Rational Mech. Anal.* **124** (1993), 43–65.

[Feller (1966)] Feller, W. An Introduction to Probability Theory and Its Applications: Volume II. New York: Wiley, 1966.

[Funaki (1998)] Funaki T., Woyczynski W.A., Interacting particle approximation for fractal Burgers equation, in *Stochastic Processes and Related Topics: In Memory of Stamatis Cambanis 1943–1995*, I. Karatzas, B.S. Rajput, M.S. Taqqu, Eds., Birkhäuser, Boston (1998), 141–166.

[Glynn (1990)] Glynn P.W., Diffusion approximations, in *Handbooks in Operations Research and Management Science, Vol. 2, Stochastic Models*, D.P. Heyman and M.J. Sobel, Eds., North Holland, New York, 1990, pp. 145–198.

[Jumarie (2001)] Jumarie, Guy. "Fractional master equation: non-standard analysis and Liouville-Riemann derivative." Chaos Solitons & Fractals 12 (2001): 2577–2587.

[Kanter (1975)] Kanter, M. (1975). Stable densities under change of scale and total variation inequalities. The Annals of Probability, **3**(4): 697–707.

[Kilbas (2006)] Kilbas, A. A., H. M. Srivastava, and Juan J. Trujillo. 2006. Theory and applications of fractional differential equations. Amsterdam: Elsevier.

[Kwapień (2006)] Kwapień, Stanisław, and W. A. WoyczyKwapieński. 1992. Random series and stochastic integrals: single and multiple. Boston: Birkhäuser.

[Kipnis (1986)] Kipnis C., Central limit theorems for infinite series of queues and applications to simple exclusion, *Ann. Probab.* **14** (1986), 397–408.

[Laskin (2003)] Laskin N., Fractional Poisson Process, Communications in Nonlinear Science and Numerical Simulation **8** (2003), 201–213.

[Liggett (1985)] Liggett T.M., *Interacting Particle Systems*, Springer-Verlag, Berlin, 1985.

[McKean (1967)] McKean H.P., Propagation of chaos for a class of nonlinear parabolic equations, in *Lecture Series in Differential Equations, VII*, Catholic University, Washington D.C., 1967, pp. 177–194.

[Margolius (1999)] Margolius B., A sample path analysis of the $M_t/M_t/c$ queue, *Queuing Systems* **31** (1999), 59–93.

[Margolius (2000)] Margolius B., Subramanian N., Woyczynski W.A., A Monte Carlo method for queuing networks which admit nonlinear diffusion approximations, (2000) (in preparation).

[Piryatinska (2005)] Piryatinska, A, A.I Saichev, and W.A Woyczynski. "Models of Anomalous Diffusion: the Subdiffusive Case." Physica A: Statistical Mechanics and Its Applications. 349 (2005): 375–420.

[Pitman (1993)] Pitman, J. 1993. Probability (1st. ed.). Springer-Verlag, New York, NY.

[Repin (2000)] Repin, O. N. and Aleksandr I. Saichev. "Fractional Poisson Law." Radiophysics and Quantum Electronics **43** (2000): 738–741.

[Rosinski (1986)] Rosinski, J, and W A. Woyczynski. "On Ito Stochastic Integration with Respect to P-Stable Motion: Inner Clock, Integrability of Sample Paths, Double and Multiple Integrals." The Annals of Probability. **14.1** (1986): 271–286.

[Saichev (1996)] Saichev, A., Woyczynski, W.A. Distributions in the Physical and Engineering Sciences. Vol I: Distributional and Fractal Calculus, Integral Transforms and Wavelets. Birkhauser, 1996.

[Smoller (1994)] Smoller J., *Shock Waves and Reaction-Diffusion Equations*, Second Edition, Berlin: Springer-Verlag, 1994.

[Srinivasan (1993)] Srinivasan R., Queues in series via interacting particle systems, *Math. Oper. Res.* **18** (1993), 39-50.

[Srinivasan (1991)] Srinivasan R., Stochastic comparison of density profiles for the road-hog process, *J. Appl. Probab.* **28** (1991), 852-863.

[Sznitman (1991)] Sznitman A.S., Topics in propagation of chaos, 166–251 in *École d'été de St. Flour, XIX - 1989*, Lecture Notes in Mathematics **1464**, Springer, Berlin, 1991.

[Uchaikin (1999)] Uchaikin, V V. "Evolution Equations for Levy Stable Processes." International Journal of Theoretical Physics. **38.9** (1999): 2377–2388.

[Uchaikin (1999)] Uchaikin, Vladimir V, and Vladimir M. Zolotarev. Chance and Stability: Stable Distributions and Their Applications. DE GRUYTER, 1999.

[Woyczynski (1998)] Woyczynski W.A., *Burgers–KPZ Turbulence — Göttingen Lectures*, Lecture Notes in Mathematics **1700**, Springer-Verlag, Berlin, 1998.

[Zheng (1995)] Zheng W., Conditional propagation of chaos and a class of quasi-linear PDE's, *Ann. Probab.* **23** (1995), 1389–1413.

Index